Images
of Earth

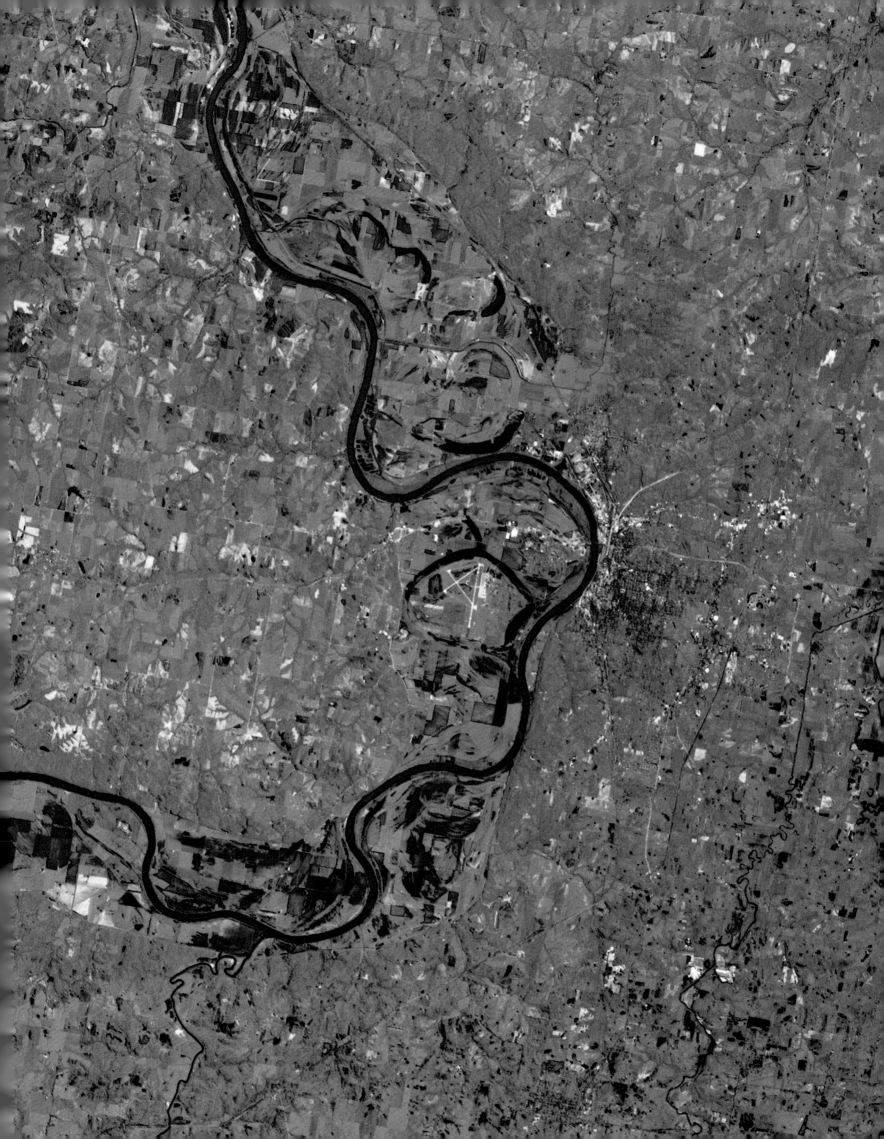

Peter Francis and Pat Jones

Images of Earth

Prentice-Hall, Inc.,
Englewood Cliffs, New Jersey 07632

Library of Congress Cataloging in Publication Data

Francis, Peter.
    Images of earth.

    Includes index.
    1. Earth—Photographs from space. I. Title.
QB637.F73 1984   550'.22'2   84-13282
ISBN 0-13-451394-0

Originally published by George Philip & Son Ltd, 12–14 Long Acre, London WC2.
This edition published by Prentice-Hall, Inc., Englewood Cliffs, New Jersey 07632.

Printed by Printers S.R.L., Trento, Italy
Bound by L.E.G.O., Vicenza, Italy

This book is available at special discount when ordered in bulk quantities. Contact Prentice-Hall, Inc., General Publishing Division, Special Sales, Englewood Cliffs, N.J. 07632.
10 9 8 7 6 5 4 3 2 1

ISBN 0-13-451394-0

Prentice-Hall International, Inc., *London*
Prentice-Hall of Australia Pty. Limited, *Sydney*
Prentice-Hall Canada Inc., *Toronto*
Prentice-Hall of India Private Limited, *New Delhi*
Prentice-Hall of Japan, Inc., *Tokyo*
Prentice-Hall of Southeast Asia Pte. Ltd., *Singapore*
Whitehall Books Limited, *Wellington, New Zealand*
Editora Prentice-Hall do Brasil Lta., *Rio de Janeiro*

Acknowledgements
Preparation of this book would have been impossible without the unique archives and facilities of the Lunar and Planetary Institute, Houston. Amongst the many individuals who helped directly or indirectly, we would particularly like to record our appreciation of the assistance and advice of: Mike Baker, Rebecca McAllister, Fran Waranius, Ron Weber, Gordon Wells, Chuck Wood, and, of course, BC, a cat with gravitas.

The image of Cerro Galan (52) was processed at the Open University and NERC, Swindon.

Note on Name Forms
The name forms used in the text are well-known English conventions where these exist. The name forms on the maps are those that are used locally and these have been included in brackets in the text where this would be helpful for reference to the maps. In the case of China, the Pinyin system of Romanization has been used on the maps. What was formerly known as the Persian Gulf, and is known in some parts of the world as the Arabian Gulf, is referred to throughout as the Gulf.

Note on Images
The image scales given in the captions are approximate. Similarly, the north point shown for each image is given to help the reader orientate the image and relate it to the map, but should not be used as an exact reference point. The dates on which each image was taken are all listed at the end of the book after the index where these are available.

TITLE-PAGE ILLUSTRATION **A Thematic Mapper image of St Joseph, Missouri, USA.**

Contents

Dedication
For all our friends at the LPI and JSC

The Belcher Islands, Hudson's Bay, Canada, taken from Shuttle 9.

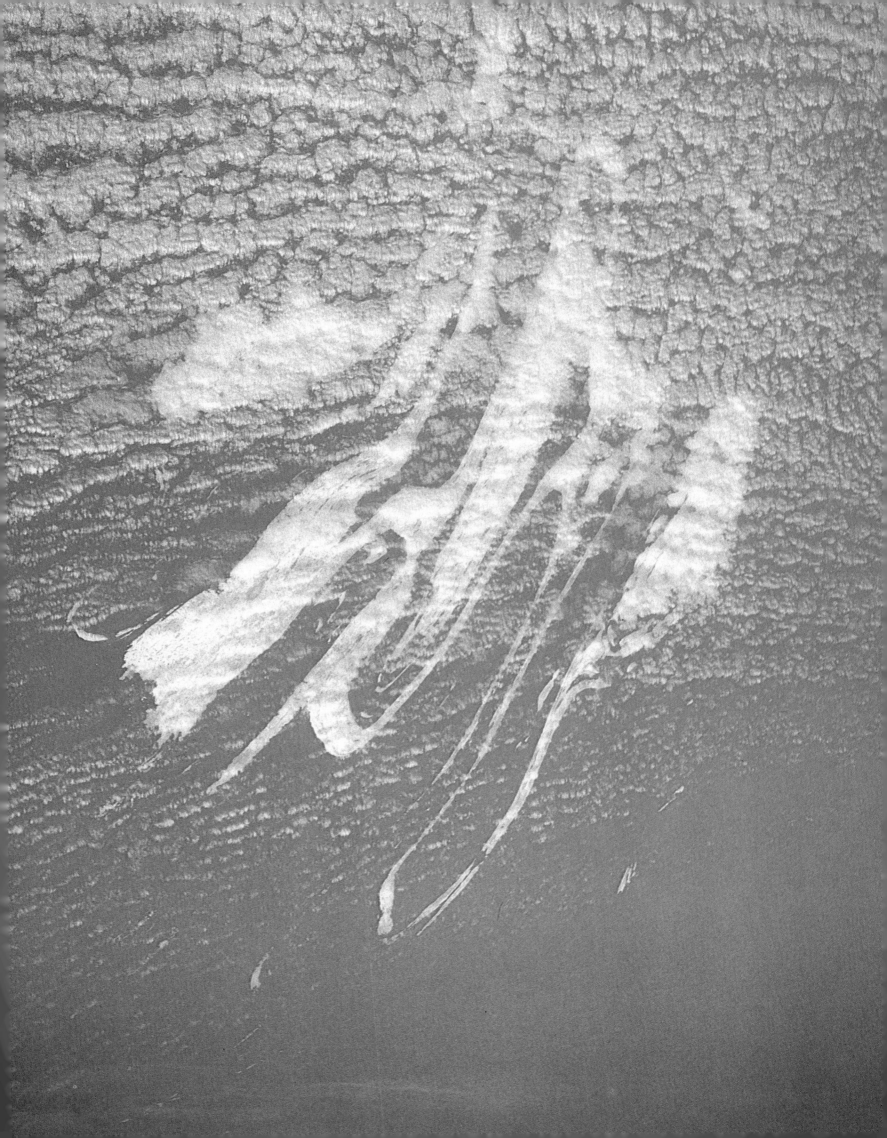

Introduction

Only a few mortals have been privileged to see with their own eyes the spectacle of the Earth shrinking and receding away from them as they head off into space: the astronauts who left the Earth for the Moon in the course of the Apollo project. All of them were humbled by the experience, and were deeply impressed by the beauty of their home planet as they saw it reduced to a small blue ball, wreathed with white clouds and brilliant against the black velvet of space. Fortunately, the rest of us have been able to share that perspective through the photographs that the astronauts took. Some of these have become among the best known of all the thousands of millions of photographs that have ever been taken, and have been reproduced in all kinds of contexts, from motifs on T shirts to illustrations on cereal packets.

It is scarcely a coincidence that people on Earth first began to become acutely aware of the fragile quality of their planet at exactly the time that these pictures became available. The concept of 'Spaceship Earth' was born, the idea of the Earth as a small, rather delicate body, endowed with a unique combination of a tenuous atmosphere and mild surface conditions, capable of supporting life. Like a spaceship, though, the Earth's resources are strictly finite and can easily be exhausted, while its environment can also be irreversibly polluted and its animal and plant life extinguished forever.

It may be many more years before man again ventures far enough out into space to recapture the Apollo perspective. Since the days of Apollo, however, a great many spacecraft, both manned and unmanned, have orbited the Earth and many of these have been launched with the specific objective of viewing the Earth from space.

Payload bay doors gaping, the Space Shuttle orbits serenely above the hazy blue of the Earth's atmosphere, with the black void of space beyond. Apart from opening to allow satellites to be deployed from the cargo hold, the payload bay doors also serve as radiators for temperature regulation. The photograph was taken from a free-flying package deployed from the Shuttle.

Sadly, many of these, perhaps the majority, have been put into orbit for military reasons. Space provides the ultimate strategic 'high ground', a lofty eyrie from which the movement of enemies can be constantly monitored. Others, however, have been launched for peaceful purposes, and have enriched the world in many ways, revolutionizing communications, weather forecasting and navigation.

In this book we illustrate two of the many different ways of looking at the Earth from space, at what the images mean, how they can be interpreted and how they can be used. Most important of all, we try to show how, by climbing a few tens or hundreds of kilometres above the atmosphere, the works of man shrink almost to vanishing point and can be seen in relation to the real, natural world of ocean and atmosphere, mountains and valleys, forests and deserts. We hope that the images from this vantage point show that, although parts of the Earth have been defiled by man, it remains a wonderfully beautiful planet, and that even man's handiwork has a fresh appeal when seen from so far away.

The two kinds of images that we have selected are ordinary astronaut photographs and electronically processed images obtained from automated Landsat satellites. In selecting images, we felt that astronaut photographs were especially important, because only an astronaut knows what the Earth really looks like from space. The simplest way for him to represent that view to us is by taking a photograph with an ordinary hand-held camera. We use many such photographs in this book, because they have a simplicity and an immediacy which is lacking in other, more sophisticated types of imagery. Furthermore, a photograph taken by an astronaut has a quality that is altogether lacking in images created automatically by a remote sensing system: the astronaut took the photograph because he wanted to, because the scene was something *he* particularly wanted to record for its beauty or interest. Remote sensing systems have no such aesthetic values.

Photographs taken on ordinary colour film with cameras that can be bought over the counter in a camera shop provide us

with a very close approximation to what it was that the astronaut saw. A photograph, however, is not reality. The colours that we see are subtly affected by the colour of the glass in the lens, by the dyes used in the film, and even by the colour of the light with which we look at the end-product. Thus, although a photograph might be the best rendering of a scene that is technically possible, it remains an image, a representation of reality.

While automated satellite imaging systems are coldly objective, always viewing the Earth vertically downwards with an unflinching stare, they have many advantages. For one thing, there is simply the question of availability. Manned space missions tend to last only a few days or weeks (although the Soviets are continually extending the duration of their missions), and for technical reasons connected with the Earth's rotation they usually orbit the Earth from west to east at relatively low latitudes. In the case of NASA's Space Shuttle, the astronauts usually view only those areas that lie less than 29° north and south of the Equator. Landsat satellites, by contrast, have been continuously in orbit since July 1972 and cover the entire Earth, with the exception of small areas near the North and South poles.

The second great advantage of Landsat satellites is that they collect data over a broader part of the electromagnetic spectrum and are not confined to the narrow rainbow band of visible radiation that our eyes can detect. Specifically, they can 'see' much further into the infra-red than we can, and thus can detect and display vividly variations in vegetation and rock patterns that we would otherwise be quite unaware of. But the data are recorded electronically. Thus, in manipulating the data from their original form, encoded on computer tape, to something *we* can see, the result is inevitably quite artificial. Colours no longer have any relation to the original. Bright red no longer means bright red, as we understand it, but may actually correspond to vegetation that looks green to our eyes. Because such manipulations are necessary to display modern remote sensed data, and because the appearance of the data displayed can be

9

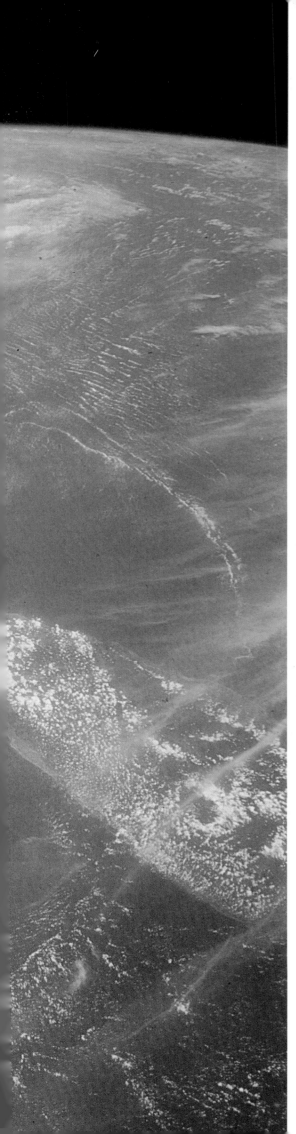

transformed with a single stroke of the computer keyboard, the products are far removed from ordinary photographs, but like photographs they are also images, representations of reality. Whether or not the end-product is visually pleasing depends largely on the whim of the computer operator, who selects the colours that we see.

Photography from the Space Shuttle

Almost all the photographs we have used were taken from the Space Shuttle. This is partly because the older Apollo and Skylab photographs are more difficult to obtain in pristine condition, but also because there is such a wealth of superb Shuttle photographs to choose from. To date there have been eleven missions and some magnificent images have been obtained on every one. Although the flights are quite short, ranging from two to twelve days, the crew is usually large enough for at least one or two members to take pictures, and the windows of the Shuttle are far superior to those on the Gemini and Apollo spacecraft, which were small and not designed to be of optical quality.

With a few exceptions, the Shuttle astronauts are all enthusiastic photographers. Indeed, many of them regard the view from the spacecraft windows as the greatest reward for their long years of training, and spend as much time as they can looking out at the Earth drifting steadily beneath them. Few people in such a position could fail to succumb to the temptation to take photographs, and the astronauts are no exception. Many have trained for more than ten years before their first flight, which in some cases is their only one. Thus, they have a very natural human instinct to record as much as possible of the greatest adventure of their lives. Apart from such personal satisfactions, photography is also an essential part of each Shuttle mission for public relations' and scientific purposes. In addition to documentation of particular activities, such as the launch of satellites carried in the Shuttle's payload bay (cargo hold), each mission also has a specific programme of Earth observations to be fitted in around the other critical parts of the mission.

Weeks before a mission, a group of specialists compiles a list of possible targets for photography. Many factors

The Indian subcontinent wreathed by cloud with the island of Sri Lanka at lower right (photographed by the Gemini astronauts in high Earth orbit, 14 September 1966).

influence the selection. In large measure, it is dictated by external constraints, such as the time of year and the exact time of launch, which controls whether it will be day or night over any particular part of the Earth's surface as the Shuttle comes overhead. Winter's low sun angle also means that some areas will be better observed at one time of year rather than another. On top of these constraints, there are also topical considerations. If a volcano is known to be erupting in some part of the world, high priority will be given to obtaining a photograph of it. If unusual weather conditions are causing flooding somewhere, this too will be targeted. After these priority targets come more routine subjects, of interest to individual specialist groups, such as desert sand-dunes, ocean wave patterns, or cloud formations. On board the Shuttle, the astronauts have a carefully programmed computer that tells them exactly when they are due to pass overhead a particular target, so that, if they are not occupied with more pressing mission objectives, they can be ready and waiting with a camera when the target comes into view. With such careful preparation, they are able to identify and locate remarkably small targets, like a single volcano amidst dozens of others in the Andes, or a lonely coral atoll in the midst of the Pacific. Inevitably, of course, there are times when clouds mask the target, or some other distraction comes along at the critical moment, but by and large the record of success is extremely good.

Technically, the equipment used is simple but effective: Hasselblad cameras fitted with a choice of 100 mm or 250 mm lenses. The Space Shuttle customarily orbits at quite a low altitude, around 280 km (175 miles), which provides an excellent view of the Earth. If the 250 mm lens is used under good conditions, features as small as 20 m (65 ft) across can be seen. Clearly, if the orbit is even closer to the Earth, then more detail will be distinguishable. Standard colour transparency film is used, but this is 70 mm wide rather than the 35 mm most of us are used to in everyday photography. During the early missions, when great attention was paid to the weight of each and every item taken on board the Shuttle, the allocation of film was rather miserly and astronauts complained that they did not have enough film to photograph everything they would have liked. In later missions, the allowance of film was more generous and accordingly more photographs were taken. Perhaps inevitably, though, this meant that the quality of pictures taken was more variable and some boring and repetitive shots resulted.

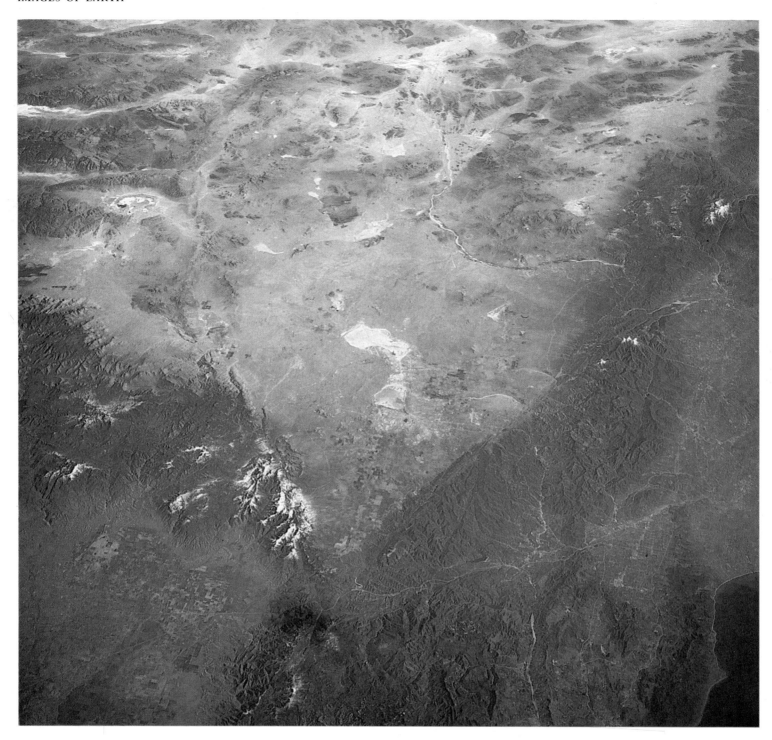

Journey's end. The white expanse of the dry lake bed at Edwards Air Force Base, California (centre) beckons astronauts as they approach the end of their mission. The consistently fine weather of the Mohave desert has ensured that all but one of the Shuttle missions has landed at Edwards, rather than in Florida, and most will continue to do so. The sprawl of Greater Los Angeles is visible on the right margin.

Although targets on Earth were carefully pre-selected, the astronauts also tended to snap anything else that caught their eye. On the first missions there was no way of recording the time at which each frame was shot, so the resulting films presented a geographic challenge of the first order. What does frame So9 . . . show? Is it part of the Inirida river in Colombia, the Sen river in Cambodia, or the Alibori river in Benin? With great skill and dedication, a handful of specialists at the Johnson Spaceflight Center in Houston has minutely examined every frame and identified nearly all of them. No easy task, given the number of pictures, the vast range of possibilities around the world to choose from, the clouds that frequently mask crucial details, and the unusual orientations that oblique views from orbit often provide. It is ingrained in us from childhood to think of maps with north at the top and it can be quite difficult to recognize even the most familiar pattern of coasts and rivers if south is at the top.

Landsat Imagery

While Shuttle photographs may offer some peculiarly puzzling perspectives, they present few problems of interpretation. Our eyes are used to looking at photographs and we do not find it difficult to appreciate that this is what such and such an area would look like from space. Landsat images are different, and some of them require a little more interpretation if they are to be comprehensible. The satellites and their orbits are designed such that they come vertically overhead every point on the Earth's surface once every eighteen days. All the images are taken looking straight down, and any geometrical distortions are cleaned up in the computer processing so that each image is displayed like a map, with north at the top. Because maps are flat, two-dimensional representations of the spherical Earth, some means has to be found of dealing with the geometric distortions implicit in flattening a sphere, and a number of map projections have been devised to overcome this particular problem. Landsat images are conventionally computed to fit a standard Mercator projection.

In an ordinary camera used on the Shuttle the film is 'panchromatic', that is, it is sensitive to all the colours of light that the eye can see, from red to violet. The Landsat sensors, however, are much more selective. In the basic multispectral scanning system there are four sets of sensors, each scanning a different part of the spectrum. One views the world at green wavelengths, one at red and two in the near or 'photographic' infra-red. The data from each set of sensors can be displayed individually, so that one looks only at the green, for example, or combined in a number of ways to produce a 'false colour' image. On the new version of Landsat satellites there is an instrument known as the Thematic Mapper which has seven sensors. These cover more of the spectral range, and include the thermal infra-red which is sensitive to temperature and can detect, for example, differences between warm and cold water. Notice that there are no sensors working in the blue part of the spectrum. This is because these wavelengths are badly absorbed and scattered in the Earth's atmosphere, so 'blue' images would look hazy and would not be a great deal of use.

By far the majority of Landsat images used in this book were made by combining data from the green, red and infra-red sensors. This might seem to suggest that the images should all look reddish green. While a few do have this colour, this is entirely fortuitous since the appearance of a false colour image depends entirely on how it is printed. Without going into technical details, data from the green sensor appears as though it were blue, data from the red sensor as though it were green and data from the infra-red sensor as

though it were red. To compound the complications, we also have to bear in mind that any individual surface reflects different wavelengths of light differently, giving it a distinctive 'spectral signature', and that no two kinds of surface have exactly the same signature. Thus, a broad swathe of trees in leaf absorbs most wavelengths except green, which is reflected, so to *our* eyes the trees appear green. But leaves reflect infra-red radiation much more efficiently than they do visible wavelengths, so they appear 'brighter' to the infra-red Landsat sensor than they do to either the green or the red sensor. The effect of this in an image is that vegetation which we 'know' is green is actually displayed as red. Many other strange effects result when wavelengths are shuffled in this way, but the only really important point to remember in looking at Landsat images is that areas which look *red* actually represent green vegetation. (On Shuttle images, green vegetation rarely shows the strong, fresh colours we are used to on the ground. It usually looks a subdued bluish grey, very similar to the colours seen from the tops of high mountains and hills. This is due to the haze-making effects of atmospheric scattering and absorption mentioned earlier.)

Landsat satellites orbit at much greater altitudes than the Shuttle, at 917 km (573 miles) above the Earth. This rather odd altitude is cleverly chosen so that the satellites keep time with the Sun. The terminator, or boundary between day and night on the Earth, marches westwards at 15° of longitude per hour, while the satellites orbit the Earth in near-polar orbits every 1 hour 43 minutes. But the Earth keeps spinning beneath the satellites, so the net effect is that their ground tracks keep shifting westwards, at about 25° per orbit. After 1 hour 43 minutes the Earth has also rotated just over 25°, with the

Landsat's near-polar orbit.

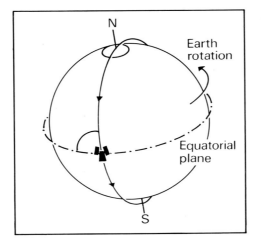

result that the ground tracks move westward at exactly the same speed as the terminator. The satellites are planned to be ahead of the terminator so that the local time beneath the satellite is about 9.30 am.

It is, of course, not as easy to see small details on the Earth from 917 km (573 miles) as it is from 280 km (175 miles). The standard Landsat sensors are able to see features no smaller than 80 m (260 ft) across, though the new Thematic Mappers have better optical systems and can see features as small as 20 m (65 ft) in width. An area 80 m across on the ground is quite large, about the size of a small city block. Many things could be crowded into a single block, of course, from people and cars to buildings and alley cats. It might seem, then, that the detail shown on a Landsat image is not really good enough, if so many small features cannot be made out. In fact, this lack of detail is more of a virtue than a problem in many cases, because often what is interesting is the *average* characteristics of an area, not the individual components. Landsat enables us to see the wood, rather than being distracted by the trees.

As everyone who has looked out of the window of an aircraft knows, altitude tends to flatten topography. Thus, only high mountains or steep slopes are really conspicuous. Since spacecraft fly so much higher than aircraft, this effect is enhanced, so it is sometimes difficult to 'read' the topography on a satellite image, particularly one taken looking vertically downwards in the standard Landsat format. When working with conventional aerial photographs, it is customary to use overlapping pairs taken from slightly different points. When viewed with a stereoscope these pairs yield a three-dimensional effect in which the topography is not only easy to read, but is also often startlingly exaggerated, so that it seems to leap up off the photograph. Much thought has been given to trying to obtain stereo cover from satellites in order to help topographic interpretations. At high latitudes there is a considerable amount of overlap between adjacent Landsat passes, so some stereo effect is available, but the system is far from satisfactory. Ultimately, a satellite remote sensing system may be developed which does provide stereo cover.

Although vertical Shuttle views present the same problems as vertical air photos, the Shuttle has the great advantage of being able to provide oblique views from low orbital altitudes, which means that large features such as mountain ranges and cloud banks photograph magnificently and are easy to interpret. If the astronauts take two or more pictures in quick succession, the pictures can be viewed in stereo and become even more impressive. For specific targets, such as an individual volcano, the astronauts can take a series of vertical views, with stereo overlap, which provides scientists on the ground with photos that can be used like ultra-high altitude air photographs, eliminating many of the problems involved in fitting together dozens of individual shots taken from conventional aircraft.

Searching through the thousands of available images to select those to use in this book was a difficult but delightful task for us. Inevitably, our greatest problem lay in the fact that we had to leave out a much greater number of images than we could ever hope to include. Apart from their visual appeal, we also tried to choose images that had some other interest, such as illustrating some aspect of history, politics or natural processes. Every single image contains such a wealth of information that one could spend several pages describing it. Rather than attempting such detailed descriptions, though, we have tried to let the images speak for themselves to a large extent, and have only highlighted or commented on those particular aspects that appealed to us. Readers with dissimilar interests or coming from different cultural backgrounds will undoubtedly see other aspects of the images and find other points of interest in them. For anyone to attempt to describe and interpret such a diverse range of images requires the impossible: that he be an expert on everything from the physics of clouds to the Treaty of Versailles and in addition have detailed knowledge of the local geography of everywhere in the atlas from Abu Dhabi to Zimbabwe. Thus, we are very conscious of our limitations. If you have local knowledge of any of the localities in the images, or if you disagree with our interpretations, we hope that you will write to us and tell us what you know. Apart from our interest in learning more about this beautiful world, this would also help us to revise and improve any subsequent editions of the book.

. . . on the sand, half sunk, a shattered visage lies . . .
And on the pedestal these words appear:
'My name is Ozymandias, King of Kings:
Look on my works, ye mighty, and despair!'
Nothing beside remains. Round the decay
Of that colossal wreck, boundless and bare
The lone and level sands stretch far away.

Percy Bysshe Shelley (1792–1822)

1 The Shaping of History

History is about people. We have records of one kind or another that extend back over five millennia and which document countless wars between nations, the rise and fall of kings, natural disasters and great human achievements. The overall impression is one of constant flux, with even the greatest rulers fading away so completely that, like Ozymandias, the only evidence of their ever having existed at all is preserved by chance in archaeological remains.

Underlying the shifting patterns of human history, however, and influencing every subtle variation, is the guiding force of the physical characteristics of the Earth; what we might call the topographic imperative. The British are, and like to think of themselves as an island nation. Yet, in the geologically recent past, Britain was joined to Europe by a broad land bridge. How different British history would have been if that bridge had not been submerged beneath the grey waters of the English Channel! Topography works on a much finer scale, too, in influencing where we live and work. The city of London grew up where it did because, in pre-Roman days, it was the lowest point on the Thames where the river could

1 Lesotho, a black nation totally encircled by white South Africa (Landsat, false colour). Image scale 1 cm = 5 km; map scale 1 cm = 20 km.

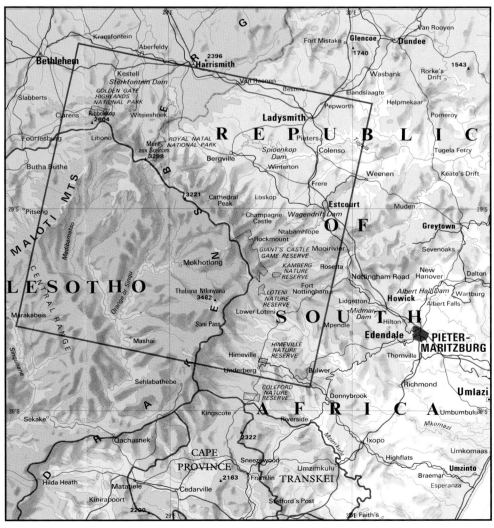

conveniently be crossed. Although technology has come a long way since then, and London has spread enormously, the great gash that the Thames estuary makes into the east coast of Britain still constitutes a formidable obstacle to travel. The Thames has indeed been bridged a few kilometres further downstream than in Roman days, but not much, and London still exists in the same place.

While the influence of topography is often straightforward, sometimes even brutally so, it also works in much more subtle ways. Here satellite imagery comes into its own, and helps us to stand far enough back from individual towns, cities or regions to be able to see how they fit into a broader topographic whole, and how the topography has affected history. One area where politics, culture and history are clearly intertwined is the Drakensberg plateau in South Africa (Fig. 1).

This great plateau dominates the whole of southern Africa. It includes the highest peaks in the southern part of the continent, which, as the image shows, are snow-capped in winter. Mont aux Sources rises to 3298 m (10,820 ft), while the highest peak of all, Thabana Ntlenyana, reaches 3482 m (11,424 ft). The plateau, much of which is above 2000 m (6000 ft), consists of a great pile of volcanic lavas which has been eroded back to form the soaring encircling escarpment. This escarpment makes a highly effective barrier between the plateau and the lowlands, and conditions in these two areas are noticeably different.

The lowlands, as the image shows, are evidently congenial. Intensive agriculture is revealed in the patchwork of fields, and there are many roads, reservoirs and towns. The largest town is Ladysmith, site of a major siege during the Boer War (1899–1902). By contrast, the highlands are bleak and chilly, and scarcely a single field can be seen. Such physical contrasts are obvious. What the image does not reveal is the political context. The cultural difference across the escarpment is as great as the physical: the lowland is white South Africa; the highland plateau is black Lesotho, an independent African kingdom totally encircled by South Africa.

How did Lesotho come to exist, a black enclave in a fundamentally hostile country? Why did the whites simply not move in there as well? Most of the answers can be found in the image.

The bleak, windswept, treeless plateau, entirely underlain by lavas, contained nothing of interest to the white colonists. The ground was too high and too cold to be suitable for any agriculture other than grazing cattle, and the lavas were completely barren of anything likely to be of economic importance. Lavas do not contain gold or diamonds. Had it been otherwise, it would not be too cynical to suggest that Lesotho would now be part of South Africa. Lesotho, in fact, has so little in the way of natural resources that fully 20 per cent of its adult males are obliged to seek employment in South Africa as migrant workers. Thus, its 'independence' is a little hollow. So dependent is Lesotho on its encircling neighbour that it has been described as a 'waif', sustained but not developed by the South African economy. It has also been described, more acerbically, as a 'hostage state'.

Today, Europeans form only a small proportion of the total population of South Africa. South America, colonized from Spain and Portugal, evolved quite differently. The bulk of the population of most South American countries is of European or mixed descent, and only in parts of Peru and Bolivia are there large numbers of Indians. The fact that the entire continent was colonized from either Spain or Portugal has led to a superficially marked degree of cultural homogeneity. Topography and situation, however, have resulted in each of the various republics developing its own distinctive national character.

Chile is almost as singular in its way as Lesotho; a shoestring of a country stretching for 4000 km (2500 miles) along the western seaboard of the continent. It is so narrow that most parts of it do not stretch across the full width of a single Landsat image (Fig. 2). As most of the length of Chile adjoins Argentina, one might suppose that the two countries would have much in common. Far from it. The apparent proximity is almost meaningless on the ground. The Andean cordillera is so great an obstacle that the frontier is only crossed by rail at two points. And while there are innumerable tracks winding across the border via remote mountain passes, there are only half-a-dozen crossing points that can be negotiated by ordinary cars or trucks.

Figures 2 and 3 straddle the Andes and reveal how sharply different Chile and Argentina are. The Chilean image is dominated by red, showing that the area is covered in green vegetation; the Argentinian image is largely barren. Chile gets the benefit of the prevailing westerly winds blowing off the Pacific, with heavy rainfall as they rise over the mountains. Argentina, by contrast, lives in the rain-shadow of the

2 The vineyard area of south central Chile, home of some of the world's finest wines (Landsat, false colour). Image scale 1 cm = 5.7 km. See p. 21 for location map.

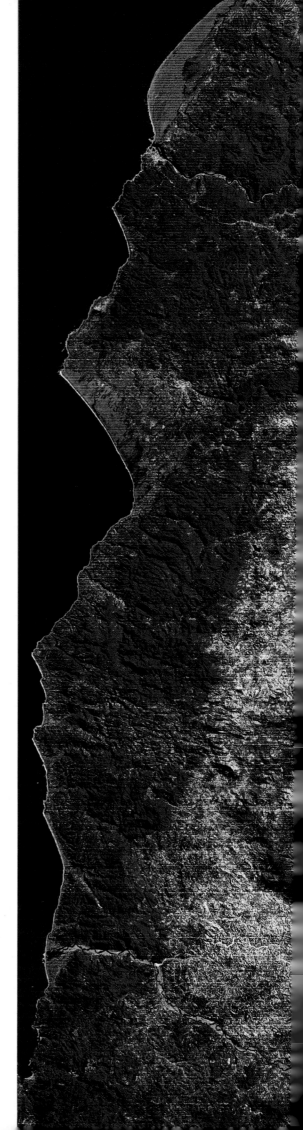

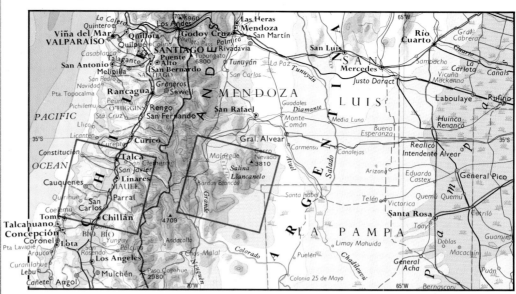

The map shows the area of Figure 2 (left) and Figure 3 (right); map scale
1 cm = 80 km.

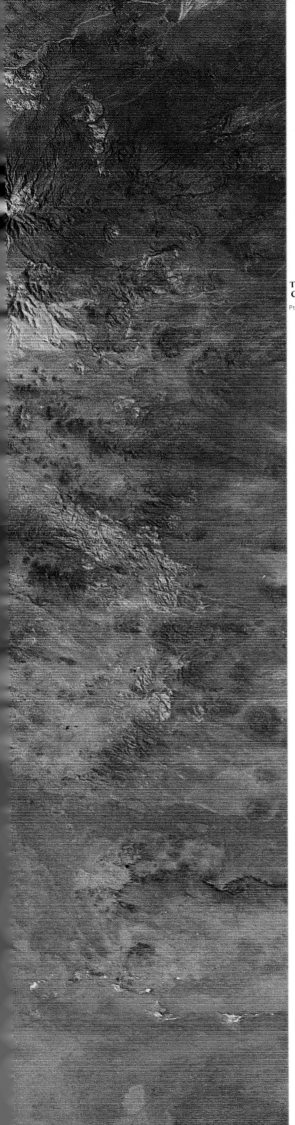

3 The volcano-studded arid pampas of west central Argentina (Landsat, false colour).
Image scale 1 cm = 5.7 km.

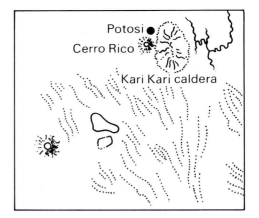

Potosi
Cerro Rico
Kari Kari caldera

N

23

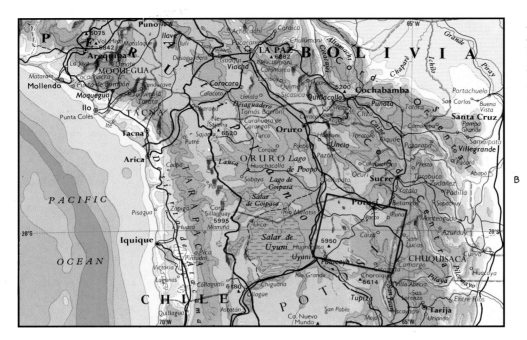

4 PREVIOUS PAGE Potosi, Kari Kari caldera and the Cerro Rico, Bolivia: source of much of the wealth of the Spanish Empire (Landsat, false colour). Image scale 1 cm = 5 km; map scale (left) 1 cm = 80 km.

Andes, and as a consequence, agriculture and lifestyles have to adapt to a much drier environment.

Most of the population of Chile lives in the Central Valley, the narrow strip of fertile land sandwiched between the minor coastal mountain range and the great wall of the Andes. This valley is well-watered by numerous great rivers draining west from the mountains. The fan-like field patterns on the rich alluvium these rivers have deposited on the flat ground of the valley floor show clearly on the image. Many of the fields visible are vineyards. The climate in the Central Valley is ideal for growing grapes, and Chileans boast that they produce the best wines in the world. Many other countries make the same claim, but when the disease known as Phylloxera devastated the classic vineyards of France, it was to Chilean vines that the French turned to replace their rootstocks.

Across the frontier in Argentina, the vast, dry plains, studded with small volcanoes, stretch away from the eastern foothills of the Andes all the way to the Atlantic. Too dry for arable farming except in a few favoured localities, the plains – or pampas – are given over to grazing cattle and sheep. The gaucho is as significant in Argentine life as cowboys are in the western USA, with the difference that gauchos still ride the pampas of Argentina, whereas the cowboys of Texas are now largely confined to expensive dude ranches, where visitors pay large sums to experience vicariously the hardships of a working life on horseback.

The dryness of the Argentine side of the mountains is revealed by the fact that there is only one river of consequence on the area of the image, the Rio Grande, while the smaller streams in the north trickle into

and dry up in the salty Lake Llancanelo. The spring-line which feeds these streams is remarkably well defined, and located along it are two clusters of large fields (red squares top left), the only visible evidence of human activity. The small town of Malargue serves as the focus for this area.

Given the vast extent of the pampas, and their suitability only for rough grazing, human settlements are few and far between. Life on the pampas centres around widely-spaced estancias, located near water supplies, where farm managers, gauchos and their families live in self-supporting groups. Towns are rare, and hard to get to, as the roads are poorly paved.

While Chile and Argentina have a distinctly European ambience, Bolivia, which adjoins their northern frontiers, has more of the flavour of Tibet (Fig. 4). Although most of its area lies in the Amazon lowlands, the historically important part of Bolivia has always been the Altiplano, and the majority of the population live at 3600 m (12,000 ft) on the cool, open plateau. Because the high plateau, like Lesotho, had little to offer colonists, the indigenous Indian population has been relatively undisturbed and remains little changed, still speaking the language of the Incas. Many of them live in small isolated estancias, herding llamas for wool, meat, and dung, which they use as fuel.

The Spanish colonists were primarily interested in silver and gold, almost to the exclusion of anything else. The devastating effect that their single-minded quest for precious metals had on the indigenous populations of the Inca cities of Peru is well known. When the Spanish reached the mines of the Cerro Rico of Potosi in 1544, legend has it that they found silver so

abundant that it could be snipped away with scissors. Such easily-won wealth triggered a massive silver rush, and Potosi became the site of frantic mining activity for the next two centuries. So many people flooded into Potosi that the population of the city in the seventeenth century exceeded 100,000 and it was easily the largest city in the western hemisphere, dwarfing anything in contemporary North America. So much silver was mined that it gave rise to a popular Spanish expression: 'worth a Potosi'. But the precious metal flooding out of Potosi caused grave problems. The large quantities of silver transported back to Spain in treasure fleets resulted in a massive inflation in the Spanish economy, from which the country never really recovered. Eventually, the easily-won silver was mined away, and the boom faded. But as the miners burrowed deeper and deeper into the Cerro Rico in pursuit of silver, they found that it was replaced by tin at deeper levels. Tin is itself a scarce metal, occurring in economic quantities only in Bolivia and Malaysia, and mining for tin continues today on the Cerro Rico, hundreds of years after the first mines were opened.

Fabulously rich mines such as those of Potosi are naturally intriguing to geologists, who want to understand how and why such concentrations of rare metals came about. The Cerro Rico is particularly interesting, because no one understood its geological setting until the advent of Landsat imagery. Many geologists, of course, had inspected the mines, and hammered at the veins, but none had stood back to consider its broader setting. Then, in 1976, two geologists poring over images of the Andes in their laboratory in Britain realized that the Kari Kari massif, south of

5 OVERLEAF Eastern Pennsylvania, USA, a region of fold mountains and heavy industry (Landsat, Thematic Mapper, false colour). Image scale 1 cm = 5 km; map scale (right) 1 cm = 25 km. The enlargement (overleaf right) shows Harrisburg and Three Mile Island with the nuclear reactor buildings appearing as white specks against the blue of the island. Scale of enlargement 1 cm = 2.5 km.

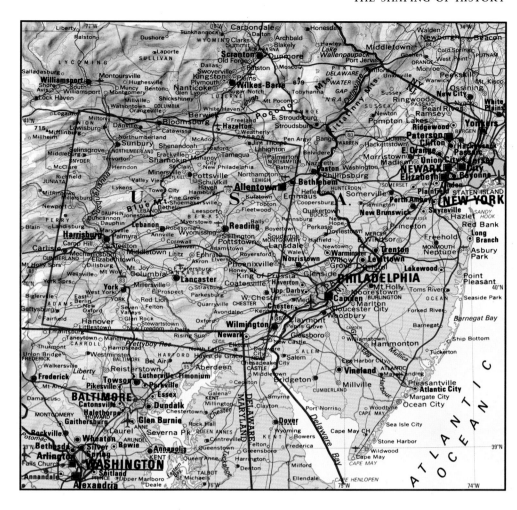

Potosi, was the centre of an enormous volcanic crater 35 km (22 miles) across, and 20 million years old. This massif, previously ignored by geologists because it was barren of mineralization, provided the key to understanding the setting of the Cerro Rico: the latter evolved from a body of rock which erupted long after the main crater formed and provided the focus for the passage of hot, mineral-rich solutions. Flowing for many hundreds or thousands of years, these hot solutions deposited enough silver to change the course of history.

The differences between North and South America's development are obvious, even in images taken from space. Figure 5 shows a large part of the state of Pennsylvania, USA. Apart from the elegant, sweeping lines of the Appalachian fold mountains which dominate the upper part of the image, and the winding sash of the Susquehanna river, the most striking feature of this image is the number of large cities visible on it. Harrisburg, York, Lancaster, Reading, Lebanon, Pottstown, Allentown and Bethlehem are all sizeable cities, visible as bluish-grey patches. Major freeways linking the cities like fine threads are also conspicuous, especially near Harrisburg, where they vault the Susquehanna on great bridges. There is also an abundance of large airports. The main runway of the Harrisburg International Airport, on the banks of the Susquehanna, is particularly easy to spot. Clearly, a dense population and much wealth are concentrated in this area.

There are many reasons why North America has evolved so differently from its southern counterpart. Some are cultural – Pennsylvania, for example, was settled by Swedish, Dutch, German, Irish and Scot-

tish colonists, who had little in common with the Hispanic settlers in South America. Pennsylvania prospered, however, not only because of the energy of its colonists, but also because it was richly endowed with natural resources.

The soil southeast of the Appalachians is rich and fertile, and supports a large agricultural industry, principally dairy farming. Notice how the tracts of fields, showing up in red, contrast with the brown of the Appalachian mountain ridges. The image was acquired in November, when the tree-covered mountains were bare and leafless, but the grasslands were still green. The fields showing up in pale grey and green tones have been ploughed, and the bare soil is exposed.

Farming, then, was the first source of wealth. The second was heavy industry. Coal is abundant in Pennsylvania, and iron ore is easily available. The first iron works was founded near Reading in 1716, and marked the start of America's industrial revolution. Harrisburg, Allentown and Bethlehem are now all major steel towns. Pennsylvania, in fact, has always been the principal steel-producing state in the USA.

Pennsylvania has continued to play a key role in the industrial evolution of the USA, but its recent contributions have been rather negative. The world recession and

the glut of cheap steel from third-world countries have blighted the old steel towns, and they are now dilapidated, depressed areas with high crime rates, high unemployment, and uncertain future prospects. During the 1970s, cheap energy from nuclear power stations was thought to provide a convenient solution to the 'energy crisis'. Many were built or planned throughout the USA. The one that was to change the whole direction of the nuclear power industry was constructed on an island, slightly south and east of Harrisburg's prominent airport (see enlargement). The site was selected so that the power station could draw on the Susquehanna for the huge volumes of cooling water that it needed. This island – Three Mile Island – is now enshrined in history. On 28 March 1979, this was the site of the worst nuclear accident in the history of the West. Although the reactor containment was not breached, and core meltdown – the so-called China syndrome – did not take place, confidence in the nuclear programme in the USA was badly shaken, and plans for new reactors have been minutely re-examined. Although there is no question of nuclear power being abandoned entirely, the lessons learnt at Three Mile Island mark a milestone in its development.

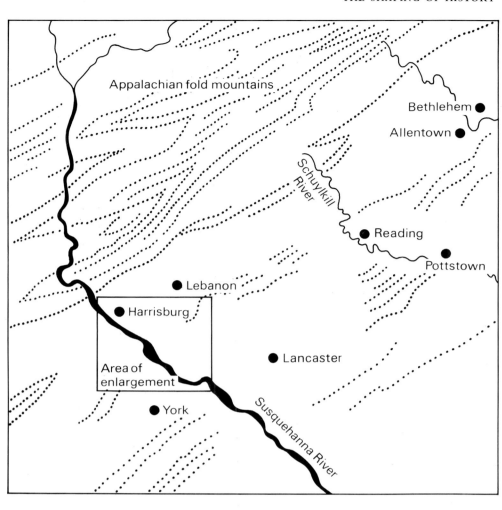

6 Hawaii, USA, showing Mauna Kea and Mauna Loa volcanoes, the highest 'mountains' on Earth (Shuttle 7, natural colour). Image scale 1 cm = 9 km; map scale 1 cm = 25 km.

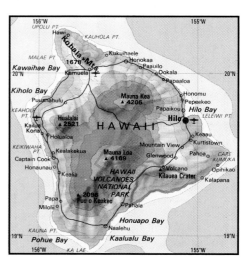

THE SHAPING OF HISTORY

Hawaii, also part of the USA, offers a total contrast to Pennsylvania. The youngest state – it became part of the Union as recently as 1959 – its development was hindered by its isolation from the mainland, and it did not become really prosperous until easy, cheap air travel became possible. Once a quiet group of islands with a rural population making a living from raising pineapples and sugarcane, the archipelago has been transformed over the last twenty years into a booming property developers' paradise.

Figure 6 shows the island of Hawaii itself, the 'Big Island'. The island is dominated by two great volcanoes, Mauna Kea and Mauna Loa, which rise some 14 km (9 miles) above the floor of the Pacific. Mauna Kea is inactive at present, and its summit at 4205 m (13,796 ft) provides the site of one of the world's most important astronomical observatory complexes. Mauna Loa is very definitely active, as the sinister streaks of dark lava from its summit testify. At the time of writing, a tongue of lava from Mauna Loa was winding down the flanks of the volcano, approaching the city of Hilo, the capital of the island. A second active volcano, Kilauea, is masked by the cloud cover, just east of Mauna Loa.

Although the volcanoes on the Big Island are the most active in the world, they are rarely violent, and usually serve as exceptionally spectacular tourist attractions. Hilo, in fact, has more to fear from the weather than the volcanoes. While Pennsylvania's wealth has been founded on its rich soil, mineral resources and communications, Hawaii's new boom depends on the weather. Hilo grew to importance on the northeast side of the island, facing the moisture-laden prevailing trade winds. As the image reveals, the cloud-shrouded northeastern slopes are well-watered, allowing the development of large sugar-cane plantations, with cattle ranching on the higher land. This agricultural economy focused on Hilo. While such activities still continue, new investment is concentrated almost exclusively on the western shores of the islands, where there is much less rain and where the sun shines much more consistently on the beaches. New hotels and resort complexes are springing up all along the coast, centred on Kona, south of Kailua, and a new airport is being developed so that flights from the mainland can land conveniently in the western sunshine. Hilo has had its day.

Much of the new development is speculative, in more ways than one. New hotels are being constructed on the surface of lava flows that are less than a hundred years old. Where lava has flowed once, it could come again. . . .

In the New World, the influence of topography on history is relatively plain and easy to read from maps and images. In the Old World, it is often more complex, not because topography has been less important, but rather because much more has happened in the long history of fluctuating national fortunes. The Shuttle image of the Strait of Gibraltar illuminates this beautifully (Fig. 7). Looking at the image without preconceived ideas, one would readily suppose (and rightly) that the Strait is of enormous strategic importance, controlling access to the Mediterranean. One might also suppose that the land on each side of the Strait would be occupied either by a single nation, exercising its sovereignty over this important sea route, or, more likely, by radically different nations on either side, as different as France and England on opposite sides of the Strait of Dover.

Reality has been much more complex. At one time, a single nation did indeed sit on the Strait, when the Arabs ruled not only North Africa, but southern Spain as well. Later, Arab influence north of the Strait waned, and Spain took over control. There was a good deal of communication across the 13-km (8-mile) Strait, of course, and as the power of the Spanish increased in the sixteenth and seventeenth centuries, a permanent Spanish settlement, founded in 1580, developed at Ceuta. The port became formally Spanish in 1688 and remains a Spanish enclave to this day. When the British emerged as the major naval power in Europe, and began to flex their colonialist muscles, they became aware of the immense strategic importance of the Strait and the value of being able to dominate it. For a few years (1662–84) they occupied Tangier, on the south shore of the Strait, but later they turned to the north shore and took Gibraltar. In 1713 this formally became a British possession, and remains so to the present day, a situation bitterly resented by the Spanish.

Oddly enough, the tiny enclave of Ceuta remained the only Spanish outpost south of the Strait. Although physically much closer to Spain, the southern shore generally came under French rather than Spanish colonial administration. Morocco, an autonomous kingdom, became part of French West Africa in 1912. In 1923 Tangier, which has always had a chequered history, became an international zone. During the thirties and forties it was a wonderful cosmopolitan city, where Arab and European influences intermixed, and where all kinds of exotic deals are reputed to have been struck. Sadly, this romantic era ended in 1956, when Morocco became independent and absorbed Tangier. Now the only romance found there is that

between pairs of pallid north Europeans, seeking cut-price sun on their package vacations.

Tiny enclaves, like Gibraltar, Ceuta and Tangier, might seem at first sight to be rather impermanent developments, since they can soon be overrun by their much larger host nations. The reverse often turns out to be the case, however, and some enclaves have existed for decades in the most unlikely circumstances. Hong Kong (Fig. 8) is a classic example. The serene silver sheen of the Zhu Jiang estuary in the image belies the tensions in the communities living on its banks. Hong Kong, the last full-blown British colony, lies on the islands on the left of the bay. Macau, a Portuguese enclave, is located on the opposite shore. Canton (Guangzhou), one of China's largest cities and trading centres, is located at the head of the bay, under the blanket of cloud.

Although the Zhu Jiang does not have quite the same strategic importance as the Strait of Gibraltar, it is none the less a vital seaway serving the heart of China, and it is easy to see why Hong Kong and Macau grew up where they did, guarding the entrance to the seaway and China's wealth. During the nineteenth century, Hong Kong became rich through trading with China, and it survived even the communist revolution. Militarily, it would have been extremely easy for China to move into Hong Kong, but it did not do so simply because it was in China's interests to have an outlet into the capitalist world. This is the chief reason why some enclaves are so enduring: they suit both parties very well.

China has given formal notice to the British government that it will take over the administration of Hong Kong when Britain's lease on the so-called New Territories runs out in 1997. Although this has caused widespread dismay in Hong Kong, it may survive as a capitalist enclave even when the British leave: the Chinese will still need their chief source of foreign currency.

While Gibraltar and Hong Kong have been the cause of many diplomatic problems, they have never caused a world war. This distinction rests with the city of Gdansk, now a part of Poland (Fig. 9), but originally a German city known as Danzig. Before World War I the German Empire was vast, and included what was then called Prussia, lying on the southern shores of the Baltic. One of the many provisions of the Treaty of Versailles which ended World War I (and is generally thought to be responsible for the start of World War II) was the separation of East Prussia from Germany by what was known as the Polish Corridor, giving land-locked Poland an outlet on the Baltic. Danzig, an important

7 The Strait of Gibraltar, gateway to the Mediterranean (Shuttle 2, natural colour). Image scale 1 cm = 8 km; map scale 1 cm = 25 km.

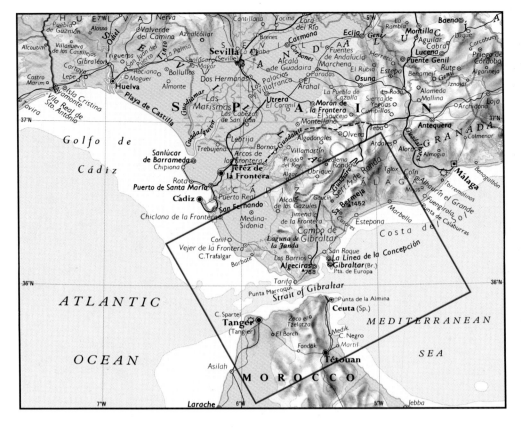

8 Hong Kong and Macau off the south China coast, last vestiges of empire (Shuttle 2, oblique view looking south, natural colour). Image scale 1 cm = 11 km; map scale 1 cm = 60 km.

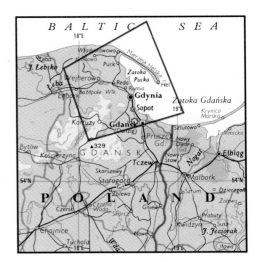

9 World War II started here: Gdynia and Gdansk (Danzig), Poland (Shuttle 9, natural colour). Image scale 1 cm = 3.3 km; map scale 1 cm = 25 km.

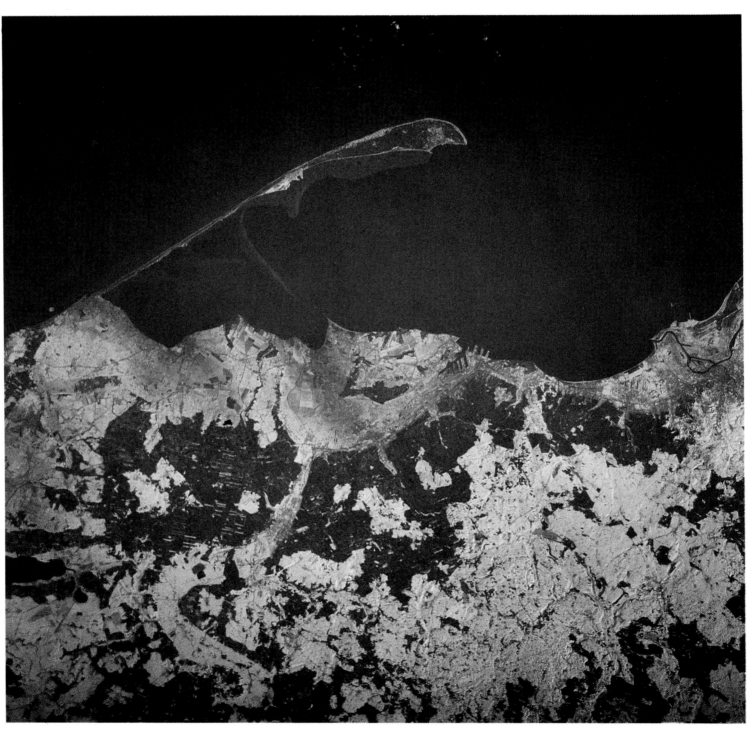

sea port, lay within the Corridor, and was established as a free city.

The German majority in Danzig was strongly pro-Nazi, and hostile to the Poles. The free city arrangement was so unsatisfactory for the Poles that they built their own port, Gdynia, a few kilometres from Danzig. The Nazi party obtained majorities in elections in Danzig in 1933 and 1935, and Hitler began to make increasingly strident demands that the city be returned to the Reich, and that the Corridor be scrapped. The Polish government refused to acquiesce to his demands, and on 1 September 1939, Hitler invaded Poland, the act that drove Britain and France to declare war.

At the end of World War II, Danzig and Prussia lay within the Soviet sphere of influence, and remain so today. Danzig (now Gdansk) is simply a Polish city. It still makes headlines, however. Gdansk and its shipyards are intimately linked with the formation of the Solidarity trade union in Poland, and its fight for a more liberal regime.

Further west along the shores of the Baltic, and forming a physical, cultural and commercial bridge between Scandinavia and Central Europe, Denmark has been a power to reckon with since the Vikings terrorized Atlantic Europe one thousand years ago. Figure 10 features the island of Sjaelland in Denmark and the southern tip of Sweden. The densely-populated urban centre of Copenhagen (København) is clearly defined. Denmark's exceptionally long coastline, fretted with inlets, has led to an economy heavily dependent on shipping and associated industries, so it is not surprising that the capital should develop around the best-situated harbour on the route between northern Scandinavia and Central Europe. The capital – Merchant's Haven in translation – lives up to its name; drawing warmth from the Gulf Stream, the city enjoys a relatively mild climate for its latitude, especially in winter. 55 per cent of the Danish population now live in urban centres with one third of all Danes making their home in greater Copenhagen. Declared a free port in 1894, Copenhagen rapidly became the most important Baltic port and its extensive docks and shipyards are clearly visible in the image. Just to emphasize the volume of sea trade conducted in this area, the equally vast docks of the Swedish port of Malmö can be seen across the Øre-sund.

Denmark is essentially a lowland nation. Its topography was mostly sculpted during the glaciation of Northern Europe in the last 2,500,000 years, when the area was scraped smooth by ice. None of the land in this image rises above 30 m (100 ft). The extent and depth of the Isefjord penetrat-

ing northern Sjaelland is a further reminder of the region's glacial history; compare the dark blue of the fjord, reflecting its depth, with the lighter shade of the shallower inland lake to the northwest.

Away from Copenhagen, almost the whole of Sjaelland is quilted with fields. The soil is Denmark's richest resource. There is very little marginal land in Denmark, and more than 70 per cent of its total area is farmed, mostly for pork and dairy products. By contrast, the country has almost no mineral wealth with the exception of oil recently discovered in the North Sea. But despite its lack of mineral resources, Denmark has for a considerable period had the highest standard of living in Europe. It is hard to identify any specific feature that has been responsible for this statistic, except for statistics themselves – there are no poor areas in Denmark to drag down national averages. The other Scandinavian countries, culturally similar to Denmark, extend far up into the Arctic and topographically are far more hostile.

Areas where ethnically or linguistically different groups co-exist are often sites of perennial, smouldering discontent, as Northern Ireland and Danzig demonstrate. Switzerland provides a remarkable contrast. Its long history of neutrality, political stability, wealth, and above all the four language groups that co-exist are all well known. But what brought about this remarkable degree of unity, in the face of so much contrary experience? Why is it that the various linguistic areas have not simply been annexed by their larger neighbours?

There are no simple answers to such questions, but the image (Fig. 11) provides some clues. Four countries are visible: Italy, in the foreground, France at top, Germany, at top right, and of course Switzerland in the centre. The large lake at the top left is Lac Léman. The Rhône valley forms a great divide across the centre of the image, making a sharp right-angled bend before entering the basin of Lac Léman. Physically, the great mountain massif of the Alps forms a formidable barrier at the heart of Europe, dividing north from south, and east from west. Switzerland is thus a perfect natural buffer state, controlling the movement of goods and people from one part of Europe to another. Such a commanding position is clearly also highly profitable, and the different linguistic groups that now occupy Switzerland have benefited greatly from working together to maximize profits from the flow of commerce, rather than discouraging trade by setting up frontiers on the mountain passes that separate one group from another.

The advantages of position apart, the high mountains, alpine pastures and narrow, deep valleys that characterize the

whole alpine region dictate a similar kind of agriculture and lifestyle, no matter what the language spoken, and thus inhabitants of French-speaking Switzerland have much in common with their German- or Italian-speaking neighbours, perhaps more so than with French speakers living in the quite different Mediterranean parts of France. Mountain people, in all parts of the world, from the Highlands of Scotland to Nepal, have always been inclined to think of themselves as superior and different from their lowland counterparts.

The Highlanders of Scotland have a particular reputation as a fiercely individualistic group. Since even the highest peaks in Scotland reach little more than 1300 m (4000 ft), the reason why this should be so would not be obvious to those unfamiliar with the land and the people. Figure 12 shows well the special nature of the Scottish Highlands.

A great diagonal slash divides the northwestern part of the Highlands from the rest; this is the Great Glen Fault, the site of many tens of kilometres of displacement. The line of the fault is marked by deep lakes aligned along the valley, of which Loch Ness is the most notorious. Whether or not it is the home of some palaeontological denizen is another matter, but what is really exceptional about the Loch is its great depth of 230 m (754 ft), with the base of the lake far below sealevel. Loch Ness was gouged out during the Ice Age, the ice sheet taking advantage of the weakness in the rocks along the fault.

The Great Glen makes a daunting barrier to travel in itself, but the image reveals innumerable other natural obstacles. There are a great many long, narrow lakes, and where there are no lakes, there are mountains. These mountains are not especially high, but, given the often filthy weather conditions in northern Scotland, they are not easily crossed, and roads are forced to follow the valleys. The coastline, too, can only be followed with difficulty as it is deeply indented by long inlets, smaller versions of the great Norwegian fjords. The image also shows clearly the many offshore islands, the most famous being Skye and Mull. None of these is very far from the mainland, but Atlantic storms and strong tidal currents sweep through the narrow channels between them. Access to the islands was a chancy, dangerous undertaking until the present century, and it is still so in some cases.

Thus, the people of the Highlands and Islands of Scotland grew up in remote, scattered communities, huddled against the seashore or on the narrow, fertile floors of valleys. With little communication with the outside world until the twentieth

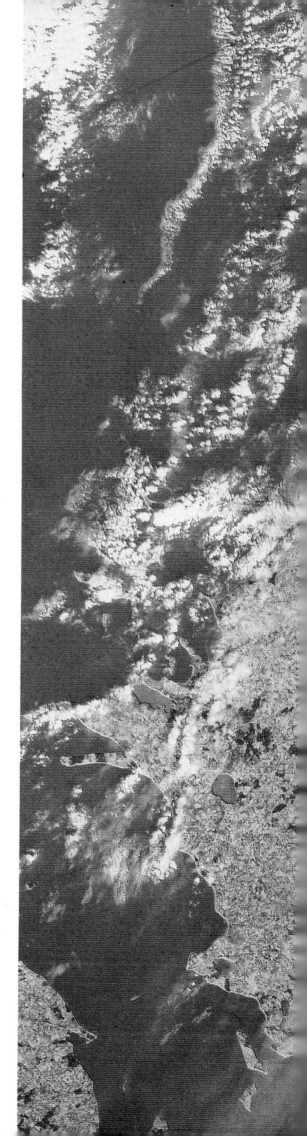

10 A highly-productive patchwork of fields and a lengthy coastline characterize Sjaelland Island, Denmark (Landsat, false colour). Image scale 1 cm = 5.5 km; map scale 1 cm = 25 km.

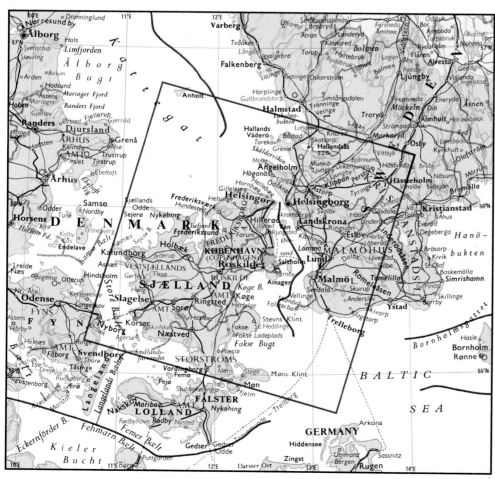

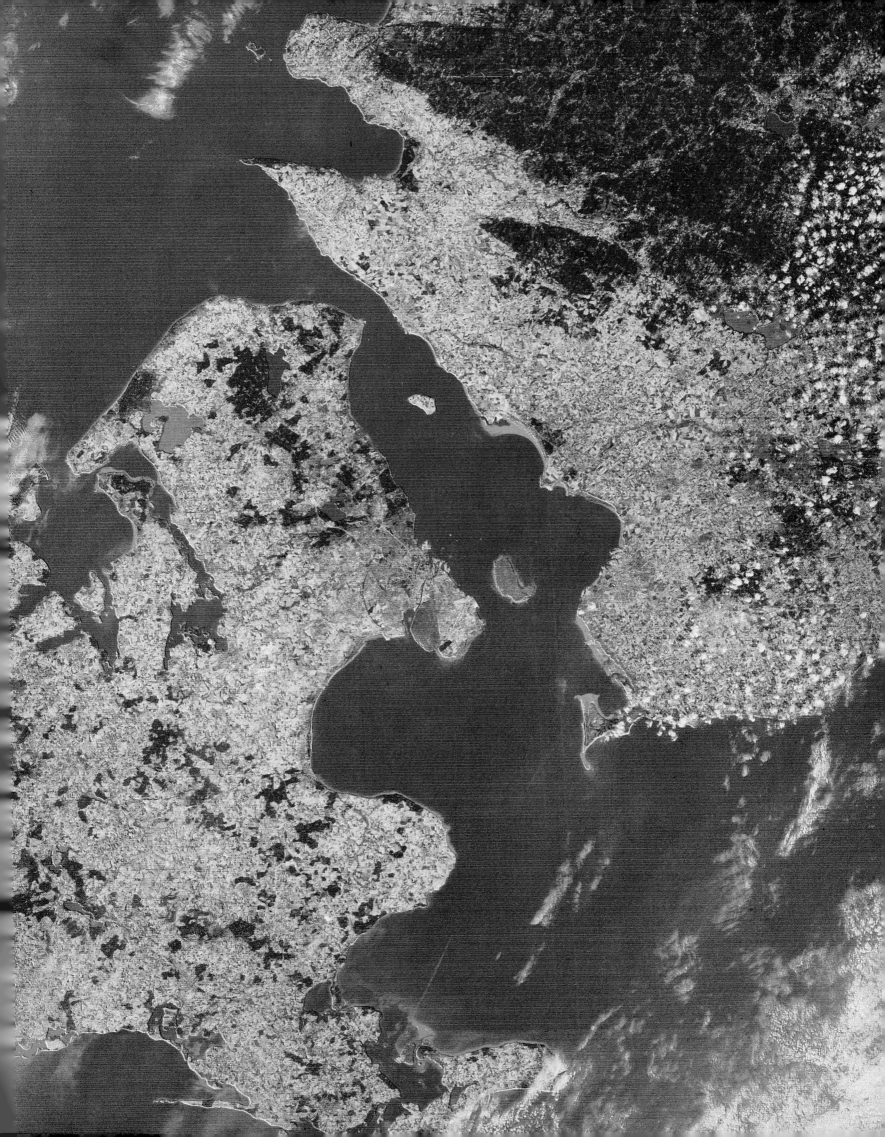

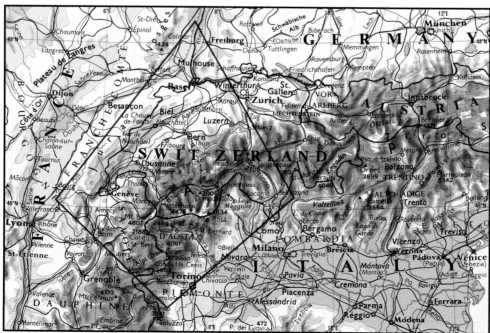

11 The mountain heart of Europe: the Swiss Alps, Lac Léman and the Rhône valley (Skylab 3, oblique view, natural colour). Image scale 1 cm = 15 km; map scale 1 cm = 50 km.

century, they retained their own language, Gaelic, which is still the first language in some communities on the outer islands. Life was always hard, since the bleak, mountainous moorland had little to offer and fishing, although rich in the waters of the Minch, was always dependent on wind and weather. The area of land suitable for agriculture has always been extremely limited – as the image shows clearly, there are only tiny cultivated patches here and there along the coast. Only around Inverness are there any extensive areas of fields.

Europe has had a long and eventful history. So too have many parts of the Middle East. The civilizations that grew up along the rivers of the Middle East, particularly the Nile and Tigris-Euphrates, are the oldest for which direct records survive. Huge areas of this part of the world are so arid and inhospitable, however, that they have been effectively uninhabited for the whole of history, apart from tiny groups of nomadic tribesmen. Figure 13 shows such an area, where the Rub al Khali juts out into the Indian Ocean to form the al Hadd Cape.

Known as the Empty Quarter, the country shown in the image actually lies in Oman, but the vast expanse of the Wahiba sand-dunes stretches unbroken into Saudi Arabia. The desert in this area historically had so little to offer that the boundaries between nation states have never been defined – dotted lines extend vaguely into the dunes on most maps. The area is not entirely uninhabited, however. Small groups of nomads, fiercely independent, live in the desert, centred on small oases, and grazing their camels and livestock on whatever pasture can be found. Some oases can be seen as small green areas between the dunes and the coastal mountains, linked by a distinct track. So clear is this track that it must be broad, probably made by successive groups of travellers moving along the same general route. Life for the nomads has always been hard. Few westerners have visited the area, one of the best known being the British writer Wilfred Thesiger. Thesiger made some epic crossings of this area of desert in the 1940s, from the Indian Ocean to the Gulf, and was enormously impressed by the tribesmen's sheer hardihood, and their skill at surviving in such a hostile environment.

Much has changed, though, since Thesiger's day, and for one simple reason: oil. Fortuitously, many of the great desert areas of the Middle East are rich in oil, and the traditional nomadic way of life has been transformed in many places, with camels giving way to four-wheel-drive Toyotas. Oil now makes up over 90 per cent of Oman's revenues, and the country's richest oil-field lies a little to the west of this image. Oil, of course, is found only where favourable geology permits it, and the coastal mountains seen on the image are made of barren rocks. Here, where modern communications are almost non-existent, life continues much as it has always done.

Although they may conceal an ocean of oil, the empty expanses of the Wahiba dunes do not reveal much about the vagaries of the oil industry. Figure 14 does, however, in quite a surprising way. This Shuttle image taken in April 1983 shows an area of the coast only 300 km (190 miles) further north than Figure 13, and is a high resolution view of the United Arab Emirates' coast on the Gulf of Oman. The image is not at first sight a very inspiring one. Look more closely, however, especially where the sun is gleaming off the water, and dozens of black specks can be seen. Each of these black specks is a supertanker, capable of carrying hundreds of thousands of tonnes of oil. Collectively, the fleet is worth tens of millions of dollars. Although a couple of tiny ports can be seen in the image (Fujayrah and Khawr al Fakkan), these are not oil ports. So why should a hugely valuable fleet of tankers be swinging idly at anchor off this rather obscure coastline?

The Arab emirates clustered together into a single nation for mutual benefit (formerly the Trucial States). Desperately poor before oil became important, the enormous oil reserves found at Abu Dhabi (Abû Zaby) have given them considerable political clout. Most of the Western World's oil comes from the Gulf, from Abu Dhabi and other producing areas further up the Gulf. The Arab-Israeli war of the 1970s produced a world-wide panic, as Arab oil suppliers shut down production. Later, the producing countries' cartel (OPEC) dramatically increased the price of oil, and thus triggered a world-wide search by consumer countries for other sources of fuel and for ways of reducing consumption. This, coupled with the world recession from about 1979 onwards, led to a glut of oil. Specifically, the reduction in consumption meant that there was a vast surplus of oil tanker capacity.

The tanker owners do not want to scrap their vessels and lose their investment, and are anxious to keep them afloat, ready for an improvement in trade. One reason for mooring them off this coastline is that there they only have a short way to steam before picking up a cargo of oil to ship to Europe or Japan. This particular mooring, however, is not an ideal one from a nautical point of view, since it is exposed to the Indian Ocean. Why not moor the vessels in the sheltered waters of the Gulf itself? Money and politics again are inextricably intertwined here. The Gulf has been in turmoil for years, and the bitter war between Iran and Iraq intermittently flares up. Threats have been made against shipping in the Gulf, and aircraft have attacked some tankers. Insurance rates in the Gulf are extremely high and so the tanker owners prudently keep their ships just outside, where they are both safe from attack, and much cheaper to insure.

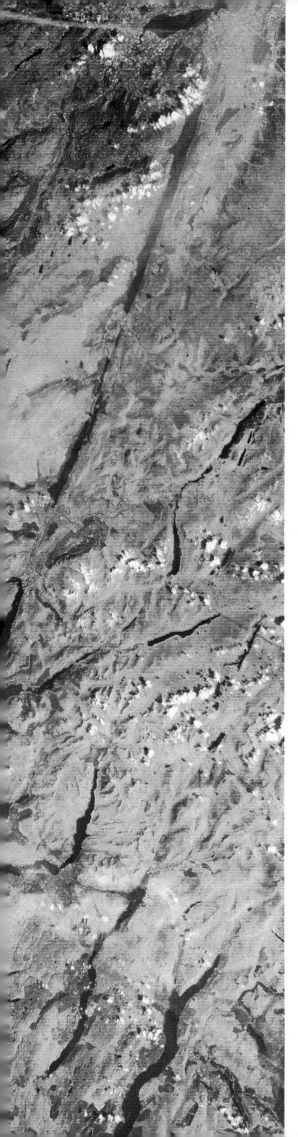

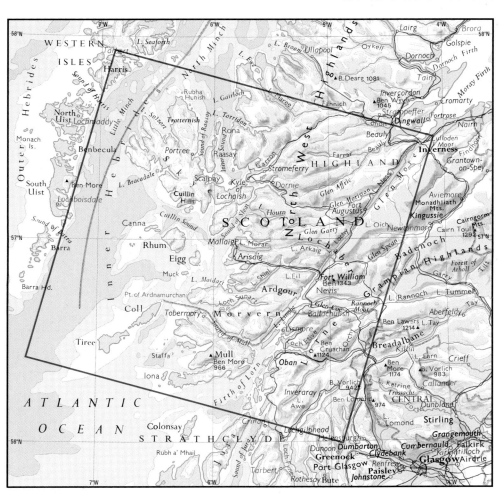

12 A hard land, mountainous and lake-strewn; the Highlands and Islands of Scotland (Landsat, false colour). Image scale 1 cm = 5.5 km; map scale 1 cm = 20 km.

13 LEFT Part of the Empty Quarter, Oman, showing the Wahiba sand-dunes: desert wastelands bypassed by the oil industry (Shuttle 9, natural colour). Image scale 1 cm = 10 km.

14 RIGHT A speculative investment: supertankers moored off the coastline of the United Arab Emirates (Shuttle 6, natural colour). Image scale 1 cm = 1.75 km.

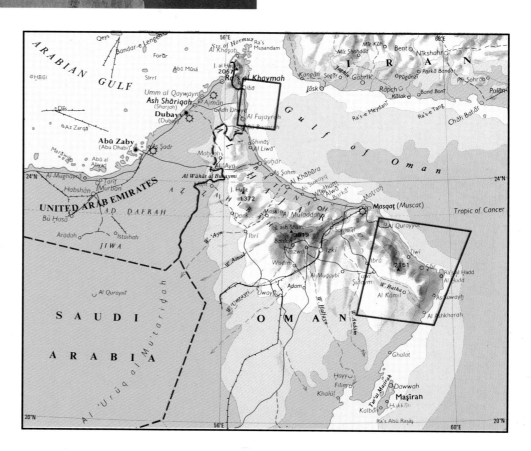

The map shows the area of Figure 13 (below) and Figure 14 (above). Map scale 1 cm = 70 km.

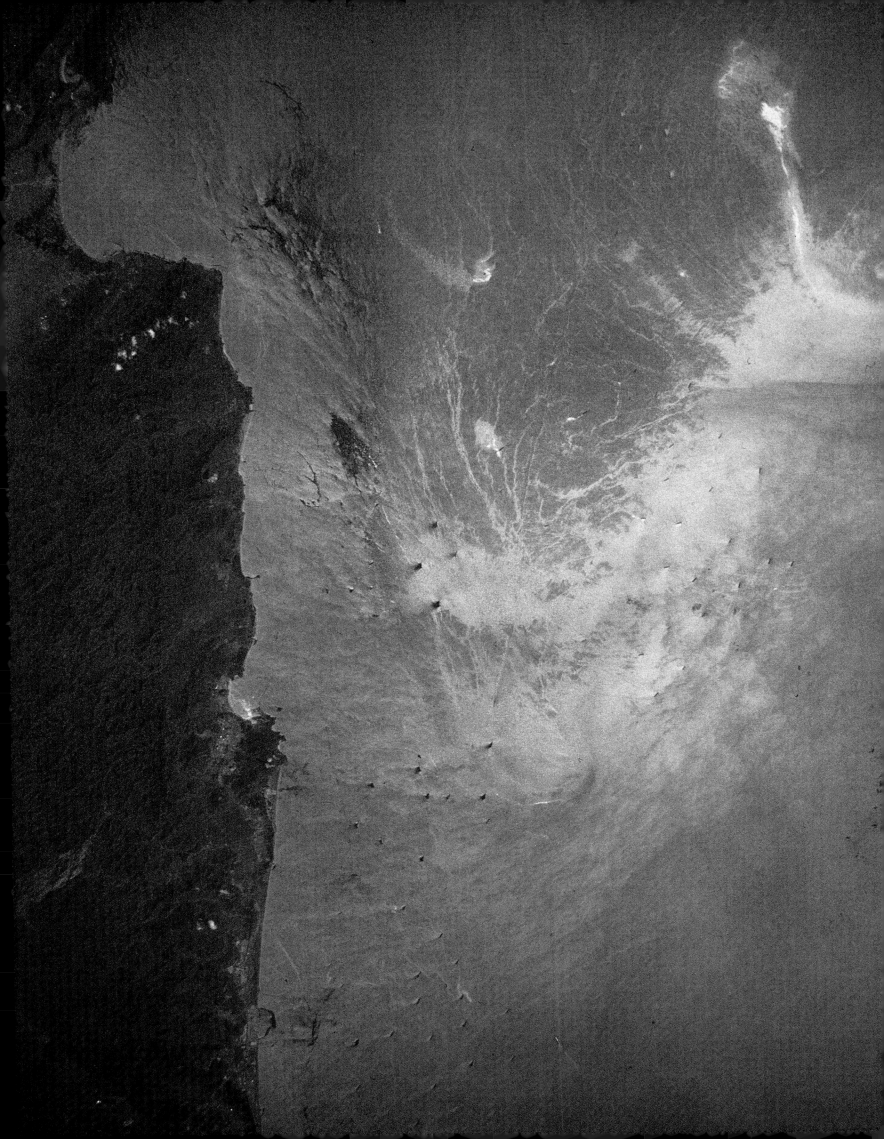

2 Nations in Conflict

From space, the Earth looks serene. While astronauts have glimpsed violent volcanic activity several times, they have never reported seeing any obvious signs of war. None the less, as they orbit they are repeatedly overflying areas where nations are at each other's throats, or where bitter civil wars are raging. Although the explosions that punctuate these struggles are mere pinpricks on the scale of our planet, and will remain invisible from space until nuclear Armageddon arrives, space imagery does provide an informative new perspective on their causes. Again, it is a question of the observer being able to distance himself far enough from the subject so that he is no longer confused by a welter of fine, usually distressingly bloody detail.

In many parts of the world, the concept of the nation-state is long established and works well – people of similar ethnic origin living within a naturally defined geographic province come to think of themselves as a natural entity. In Africa, however, the concept breaks down badly, because the European colonizing powers simply ignored tribal boundaries when they carved up the continent to suit themselves and their imperialist ambitions. Somalia, occupying the Horn of Africa (Fig. 15), is a spectacular but tragic example. The boundaries of the present Somali Democratic Republic, popularly known as Somalia, do not in any sense define a natural geographic or ethnic entity. Its present territory is made up of former British and Italian colonies and amalgamating these caused considerable problems when the country became independent in 1960. But the situation is still more complex. There is a distinct Somali people, with their own very long-established language and customs, who formed a nation long before the arrival of the Europeans. The Somalis now find themselves living not only in the present Republic, but also in the neighbouring countries of Kenya, Ethiopia and Djibouti. Given that the Somalis are largely nomadic, with little interest or regard for politician's lines drawn on maps, the present boundaries of the country are not very meaningful. The western frontier with the Ogaden area of Ethiopia has been the scene of particularly bitter fighting, with deep raids by the Ethiopian army into Somali territory taking place since 1977. As recently as January 1984, 34 school-children died in an Ethiopian airstrike against the border town of Borama.

15 **A blistering sun-burned plateau and a nation torn apart: northern Somalia on the Gulf of Aden (Shuttle 2, natural colour). Image scale 1 cm = 5 km; map scale 1 cm = 80 km.**

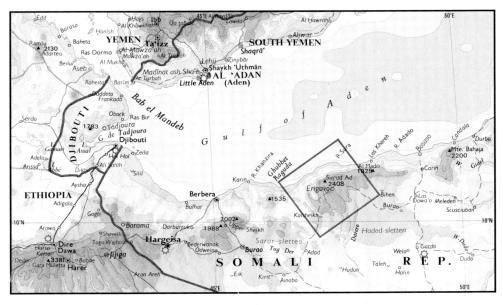

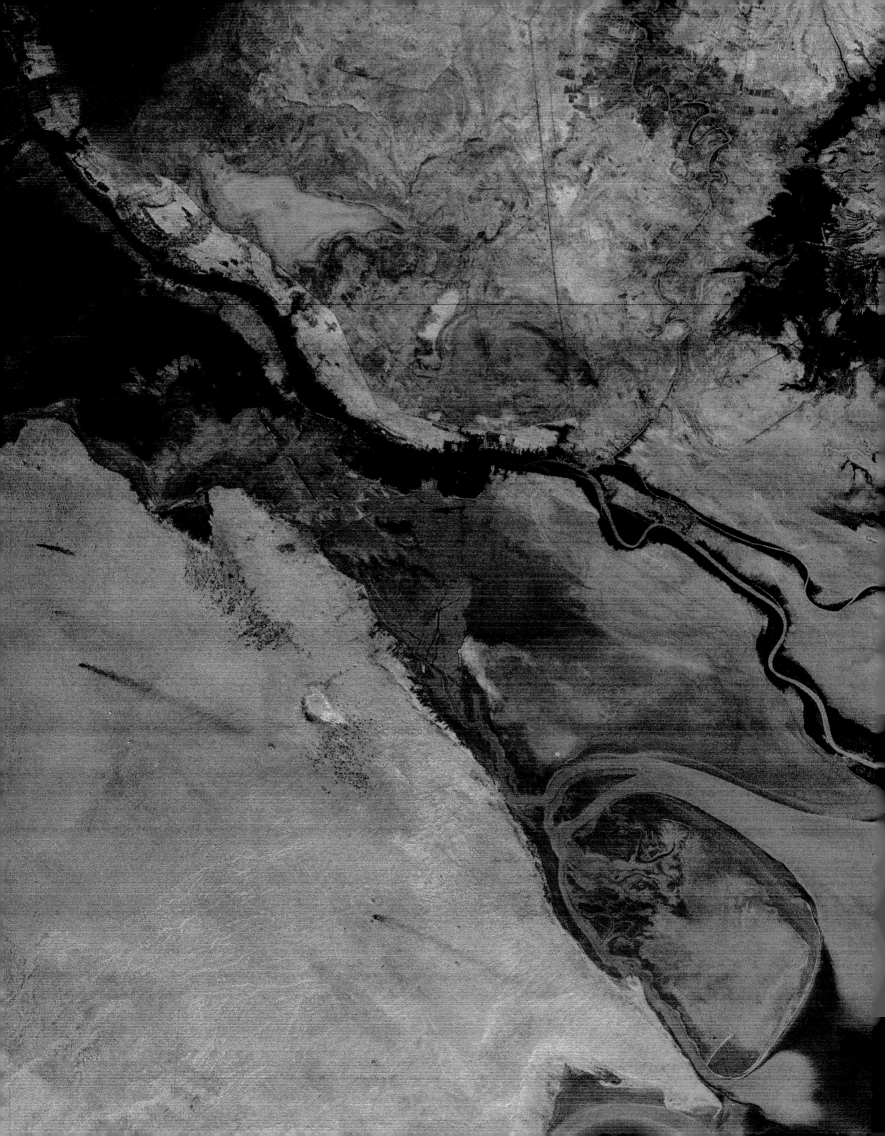

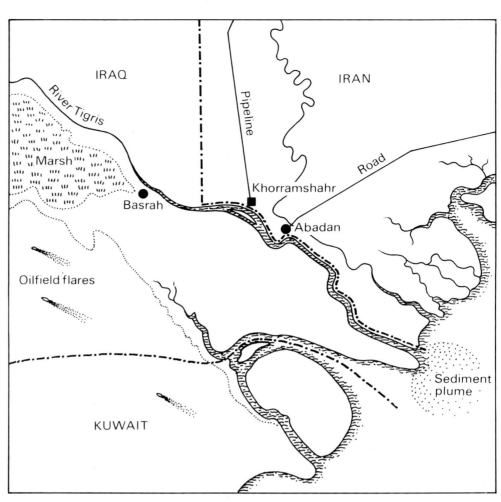

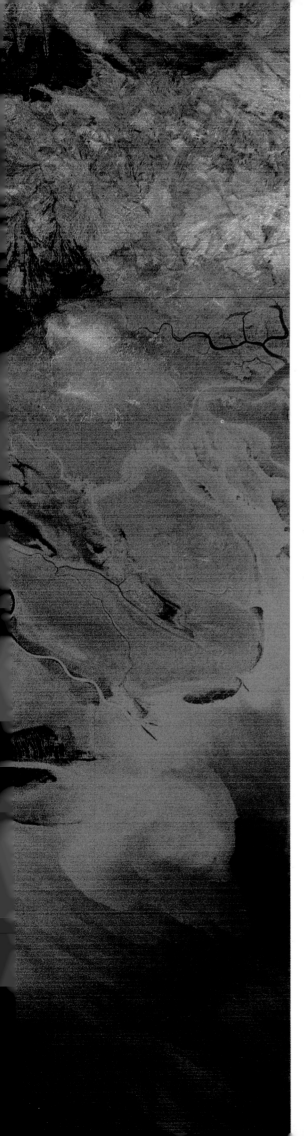

16 The Tigris river and the head of the Gulf, Middle-Eastern tinderbox (Landsat, false colour). Image scale 1 cm = 5.5 km; map scale 1 cm = 70 km.

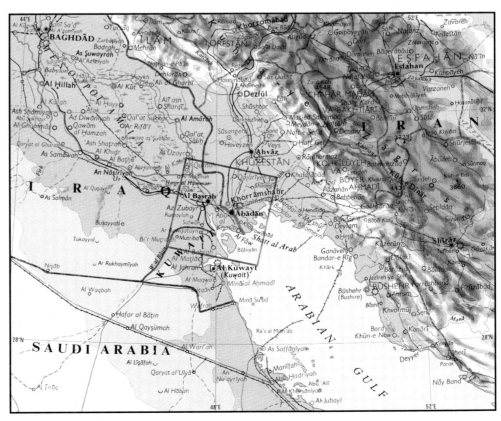

Such territorial squabbling might seem a little less pointless and tragic if there were something worth fighting over. As the image reveals, however, the Somali Republic is a desperately poor country, with a hot climate and harsh, desert terrain. The coastal strip is particularly depressing, since although it is hot and humid, rain rarely falls.

Inland from the coast, the land rises in stepped topography, well displayed in the image, to a highland plateau, in which the small town of Erigavo can be seen at the centre of a cobweb of faint tracks radiating away from it. At a height of over 1500 m (5000 ft), the climate in Erigavo is much pleasanter than on the coast, and frosts sometimes occur. Since it is only 10° from the Equator, however, the sun can be blistering, and only hardy scrub can survive – the province is known as Gubar, 'burned'. Any area of grazing in this environment can only support livestock for a short time, so the nomadic Somalis were accustomed to move around in search of fresh pastures. Even before colonial days, there were frequent feuds between rival groups over access to the best grazing. Recently, construction of irrigation schemes has helped to reduce the historic feuding over grazing rights, but the larger territorial problem smoulders on. In itself, the Somali Republic is a viable entity, but the Somalis living in Ethiopia will continue to be strangers in an alien land.

A far more serious and even more pointless war rages between Iran and Iraq, centred at the head of the Gulf. This long drawn out conflict plumbed fresh depths of horror when it became the first war since World War I to involve the use of chemical weapons. The area being fought over lies at the mouth of the Tigris river, in the flat, marshy delta lands that the river has built out into the Gulf (Fig. 16). The delta region is one of major political and economic significance since three countries meet there: Iran, Iraq and Kuwait. All are major oil-producing countries, supplying Europe and Japan with much of their oil needs. Dark plumes from massive oil-field flares can be seen in the lower left of the image, where surplus gas from the oil-fields is being flared off. A number of major towns are visible: Abadan and Basrah in Iraq, and Khorramshahr in Iran, with roads and oil pipelines leading in ruler-straight lines from them. The dense black areas on the image are waterlogged marshland.

Both Iran and Iraq have suffered severe and tragic losses through their military enterprises, and their oil revenues have been affected by the war, which seems to have been caused by a mixture of religious fervour, political opportunism, and personal rivalry. Technically, Iraq started the war by invading Iranian territory, but students of the area suggest that this was a pre-emptive strike, to prevent the Iranians attacking first.... At the heart of the matter seems to be a personal antipathy between President Hussein of Iraq and the Islamic religious zealot Ayatollah Khomeini of Iran, and whatever the original justification for the war, no matter how tenuous, neither can now stop it without losing face.

Most of the fighting has been centred around the towns along the Tigris, with the Iranians attempting to breach the vital Iraqi supply road from Baghdad (off the image) to Basrah, and the Iraqis attempting to take the oil-refining town of Khorramshahr. The rest of the world has been forced to take notice of this bitter war since Iraq and Iran between them are major suppliers of oil to the West. Japan, for example, has drawn up to 40 per cent of its oil from this region. Iran in particular has the noose in its hands, as Figure 17 shows. The jagged headland at the bottom of the picture is the Musandam Peninsula, part of Oman; the land at the top is part of the Iranian coastline, and the seaway between them is the Strait of Hormuz. Since the Strait is only 55 km (34 miles) wide at its narrowest point, it would be extremely easy for the Iranians to sever access to the oil ports of the Gulf by simply laying a minefield across it. This would simultaneously put pressure on Iraq and present a severe threat to western consumers. Although Iran has threatened to close the Strait, it has not actually done so, probably because the *threat* of closure is of sufficient strategic value and because keeping the Strait open means that the oil revenues will continue to come in.

In May of 1984, tensions in the area increased, as both sides started attacking tankers moving up and down the Gulf. Initially, Iraq seemed to be principally involved, attacking ships leaving the Iranian oil terminal on Kharg Island, but later Iran seemed unable to resist responding. The attacks had immediate world-wide repercussions, as the merchant fleets of the world began to fear what might happen if they brought their tankers into the Gulf.

Oil markets around the world became intensely uneasy and insurance rates for tankers visiting the Gulf, already high, soared to astronomic levels. Political analysts started speculating about a millennial catastrophe, since the Gulf lies just as much in the sphere of interest of the USSR as it does that of the USA. While the Soviets have no reason to wish the Strait of Hormuz closed, it would be unlikely that they would accept a heavy US military involvement so close to their own territory. (Their satellite state, Afghanistan, lies only

a few hundred kilometres inland from the Strait.) While the immediate crisis had not been resolved at the time of writing, Figure 17 is exceptionally interesting because it illuminates some of the environmental consequences of the Gulf war.

The heavy traffic of oil-tankers up and down the Gulf means that pollution by accidental (or deliberate) oil spills is inevitable, and some evidence of this can be seen in the surface slicks which are emphasized by the sun's sheen on the water. In the sunglint, the track of a large tanker is visible, its wake revealed by the churned up water and slick.

Deliberately attacking oil installations or sinking supertankers is, of course, a recipe for disastrous pollution. Sadly, this has already happened in the Gulf. A massive spill took place in 1983, partly caused by a ship colliding accidentally with an Iranian oil rig, but exacerbated by an Iraqi attack. Capping a damaged well would normally have taken ten to twelve days, but because of the fighting in the area technicians could not get near the well, and the oil continued to flow, at about 5000 barrels a day. A massive oil slick spread far down the Gulf, greatly exceeding previous spills in magnitude, and causing damage to marine life that is as yet completely unquantifiable. Although there was nothing that could be done to stop the flow of oil, satellite images were used to record its daily progress.

Every additional tanker that is sunk or set on fire in the Gulf means yet further pollution. Each event will be plain enough to the orbiting satellites, but having such an abundance of information merely compounds the frustration environmentalists feel at being impotent to prevent such acts of gross vandalism.

Simple, old-fashioned territorial expansionism underlies the conflict in the area illustrated in Figure 18, which covers the frontier between Tibet and the Kashmir region of India. The whole of the area lies above 4000 m (13,000 ft), and although the scenery is spectacular, there is even less to sustain life than in the scorched highlands of the Somali Republic. So high is the terrain, and so cold and dry is the climate that only tough, spikey grasses can survive. Although the snow-capped mountains suggest heavy precipitation, the Tibetan plateau is actually arid, and the snow cover is very thin. There are some small glaciers present, but there is nothing like the immense accumulation of ice that is present on the mountains of the Himalayas, only a short way to the south and west, which get the benefit of the monsoons. At the top right of the image is a small lake, whose surface texture shows that it is frozen over, visual confirmation of

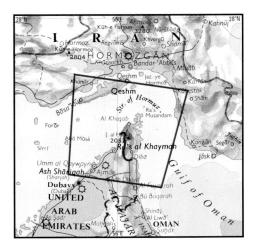

17 The Strait of Hormuz, the strategically vital entry to the Gulf (Shuttle 4, natural colour). Image scale 1 cm = 10 km; map scale 1 cm = 70 km.

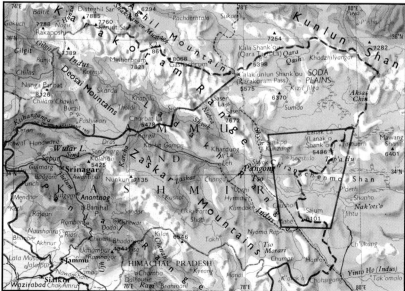

18 Pangong Tso, a lake divided by the frontier between India and China and scene of the recent conflict between them (Shuttle 9, oblique view, natural colour). Image scale 1 cm = 6 km; map scale 1 cm = 60 km.

how bitterly cold the plateau is. Before satellite images became available, cartographers had considerable difficulty mapping such remote parts of the world and the lakes shown on the map do not actually reflect what the image shows.

Deservedly known as the Roof of the World, the desolate Tibetan plateau is almost completely uninhabited, uninhabitable, and devoid of any known mineral resources. None the less, it is still apparently worth fighting over. The nominal frontier between Chinese-administered Tibet and India runs through the narrowest part of the long thin lake in the middle of the image, the Pangong Tso. Or rather, this was where the border was supposed to run. Fighting took place between India and China along this extraordinarily remote frontier during the 1970s. The conflict drew international attention for a while, but eventually petered out into a stalemate which left Chinese forces in control of over 18,000 sq km (7000 sq miles) of mountainous plateau that had originally been Indian. The Chinese claimed that the area they had overrun was theirs by historical right, an argument they also used when they occupied Tibet in 1951. The demarcation of the boundary was not resolved when the fighting ceased, and neither side seems likely to abandon its claims. Since the 1970s, conditions within China have changed substantially, with Mao and the Cultural Revolution fading into the past, and there have been no fresh territorial initiatives on the Tibet frontier. The Chinese have concentrated instead on trying to develop Tibet, and there is evidence for this in the image which shows some new roads, especially near the frozen lake.

Other conflicts clearly have an economic rather than a territorial basis. In South America, Chile and Bolivia have been bitterly hostile to one another for the last century in a dispute that is clearly related to the ownership of valuable resources.

Figure 19 illustrates compellingly the major reason for Bolivia's antipathy towards Chile: a century ago, *all* of the area in the image that is now Chilean formed part of Bolivia, and was lost by conquest to Chile in the disastrous (for Bolivia) War of the Pacific (1879–84). Originally, Bolivia owned the entire Pacific coastline from Arica southward to Antofagasta, a vast tract of the most arid desert in the world. As the image shows, the area really is remarkably barren, with red ribbons of vegetation limited to some of the river valleys – it is possible to drive for hundreds of kilometres without seeing a single blade of grass. But the visitor does not easily forget the grand, sweeping vistas, and the great gorges near the top of the image are as deep and wide as the Grand Canyon, though not as steep.

Nations do not go to war, however, for the sake of the view. The Atacama desert, precisely because of its extreme dryness, contained two natural resources of enormous importance. The first of these, which initially seems slightly improbable, was bird droppings – guano. Millions of sea-birds, feeding on the rich sea food of the Humboldt current, roosted on the rocky islets and ledges along the coast. Their droppings, never washed away by rain, accumulated over the millennia to form thick, easily quarried piles which were eagerly sought by European entrepreneurs before the days of artificial fertilizers. A single cargo of Bolivian guano, shipped to Europe by schooner, could make a fortune for an astute owner.

Second, and of much greater strategic importance, the Atacama was a major source of nitrates. Nitrates are also important fertilizers, but potassium nitrate, saltpetre, is an essential component of gunpowder. Sodium nitrate, so called Chile saltpetre, is the only large scale *natural* source of the nitrate, and it is found in workable quantities *only* in the coastal parts of the Atacama desert. Even now, the details of the natural processes which lead to the formation of natural nitrates are not fully known, but it is certain that a truly exceptional arid environment is required, since nitrates are extremely soluble in water. During the nineteenth century, as the Industrial Revolution swept through Europe and North America, the demand for nitrates boomed and boomed.

A vast country, larger than France, with a population that is still only about 6 million, Bolivia has always suffered from chronic underpopulation and all that implies in terms of lack of resources. As the nitrate boom swept through the Atacama, Bolivia was quite unable to take advantage of it, lacking the human or financial resources to invest. Most of the development was undertaken by British, Chilean and North American capitalists. Anxious to obtain direct control of the nitrate deserts, Chile invaded in 1879 and took over the entire area. Bolivia formed an alliance with Peru to try and hold off the Chileans, but in May 1880 Chile was able to defeat a combined Bolivian/Peruvian force at the Battle of Tacna, ending effective Bolivian resistance and completely cutting off Bolivia from the sea. Ultimately, the triumphant Chilean forces pushed their way up through the desert as far north as Lima. A definitive peace treaty was signed in 1904, when Bolivia conceded her vast Pacific territories to Chile. In partial compensation, Chile agreed to provide for the construction of two railways to give Bolivia some limited access to the seaports on the Pacific. One, from Arica to La Paz, runs just south of the Peruvian frontier; the other links La Paz with Antofagasta, a few hundred kilometres further south along the coast.

Civil wars are probably more common today than wars between nations. The 'banana republics' of Central America, as they are often contemptuously termed, have enduring reputations for political instability which at present are well-deserved. But why? Part of the reason may be that the region is one of the most physically unstable in the world. The chain of volcanoes between Lake Managua and the Gulf of Fonseca in the image of Nicaragua (Fig. 20) underscores the natural violence of the area. The large crater (Cosiguina) on the south shore of the Gulf of Fonseca was formed in a great eruption in 1835, one of the most explosive in recent history. The other cones, with smaller craters, erupt frequently, but less violently. Areas of fresh ash and lava flows show up as dark grey or black patches around the summit craters. The city of Managua, capital of Nicaragua, is the grey area on the shore of Lake Managua, on the bottom right margin of the image. It has had a tragic history. Ruined first by flood in 1876, it was then devastated by a major arsenal explosion in 1902, wrecked by civil war in 1912, heavily damaged by earthquake and fire in 1931, and, most recently, almost completely wiped out by a great earthquake in December 1972.

Apart from these natural disasters, Nicaragua has also had to put up with no less than nine constitutions between 1838 and 1972. Few of the presidents had much to recommend themselves, many simply running Nicaragua as a kind of extension of their business interests. This trait was nowhere better displayed than in the last president, Anastasio Somoza. He was the third member of his family to hold the office since his father assumed power in 1936 and founded the Somoza dynasty. Somoza had a controlling interest in many of Nicaragua's largest concerns, from the national airline downwards.

A corrupt, despotic presidency was not the only problem that Nicaragua faced, however. The fundamental cause of deep-seated unrest which was eventually to lead to the overthrow of Somoza by the Sandinista guerillas (named after Sandino, an early rebel leader) was the system of land tenure that the Spanish conquistadors left behind. Most of the useful agricultural land was held by a very small number of wealthy families; the peasants who formed the bulk of the population had very little. For the sake of appearances, and especially

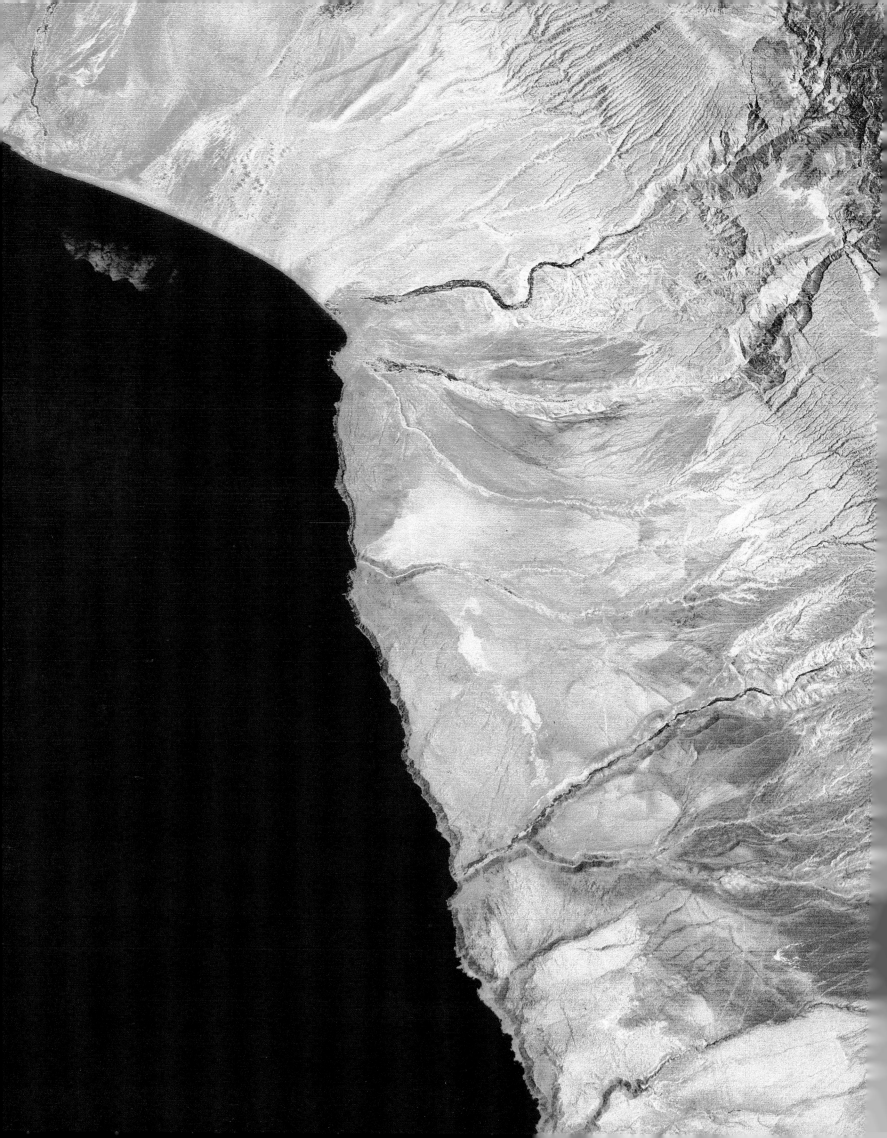

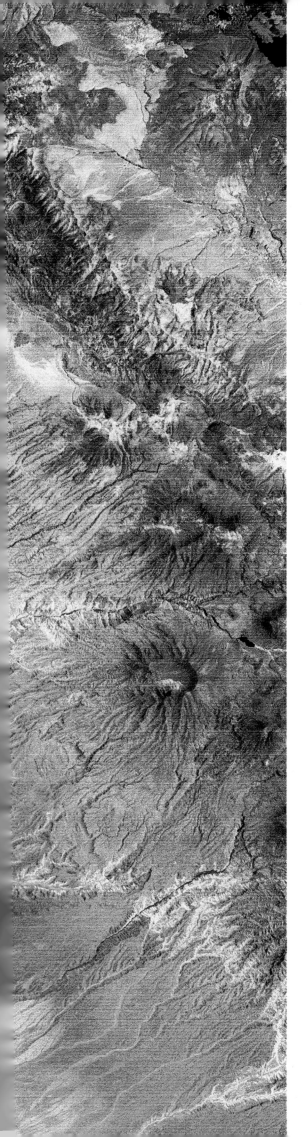

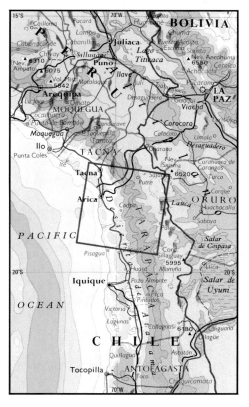

for the benefit of the USA, the country was nominally democratic, with universal franchise, but such elections as were held were a travesty with naked rigging of the ballots by the government. The Somoza government also controlled the trade unions, so the mass of the people had little opportunity to express themselves.

Although now much reviled by right-wing groups in North America because of its pro-Soviet orientation, the Sandinista movement was initially a genuinely popular movement, commanding wide support. Indeed, it could not have been so successful so quickly if it had been otherwise. The nature of the countryside itself also played an important part in the movement's success. One of the Spanish conquistadors said of Central America that its topography was like that of a crumpled sheet of paper. And so it is, as the inland area forming the majority of the image reveals. Communications are difficult, settlements are widely scattered, and remote mountain valleys offer ideal hideouts for guerilla bands. Given the vast areas that the guerillas had at their disposal, it was impossible for the central government to retain control of the more remote regions, and eventually they attempted to retain only the towns. One by

19 The barren but mineral-rich Atacama, driest desert in the world and cause of a bitter dispute between Chile and Bolivia (Landsat, false colour). Image scale 1 cm = 5.5 km; map scale 1 cm = 80 km.

one, they lost these also.

Somoza was finally deposed in July 1979, and a Sandinista regime took over. Ironically, the Sandinistas are now confronted by the same topographic problems that helped them to come to power: anti-Sandinista guerillas, operating with CIA support, have little difficulty in slipping over the frontier from Honduras, making hit-and-run attacks on towns and bridges, and then melting back into the hills.

The image also reveals another aspect of the pressure on land and resources that is characteristic of much of Central America. The vegetation on the volcanic cones is a darker red than vegetation on the lower areas. The darker tones show areas where the primary forest cover survives on the steep slopes of the volcanoes, the paler colours are secondary growths or cleared for fields. Pressure on the primary forest is immense, with demands both for firewood and extension of arable lands. Throughout Central America, the remnants of virgin forest cover are shrinking steadily upwards, and their continued existence is in jeopardy. The high, isolated cones have often developed distinctive animal and plant species, and many of these are at risk.

* * *

The two specks that make up the dry land of Midway atoll (Fig. 21) have an area of only 5 sq km (2 sq miles) and are mere dots in the vastness of the Pacific Ocean. Yet they have probably had a more decisive, direct influence on world history than any island a hundred times larger. Located 2080 km (1300 miles) from its nearest neighbours in the Hawaiian island group, Midway atoll, as its name suggests, is a natural half-way point in the watery expanse of the west Pacific.

The reef encircling the atoll has a circumference of about 24 km (15 miles), and encloses a magnificent turquoise lagoon. In two places, the reef breaks the surface to form the minute islands: Sand and Eastern. The islands were discovered in 1859, and annexed to the USA in 1867. In 1905, they first became of direct practical importance when a submarine telegraph cable was laid to link Hawaii with the Philippines, and an intermediate station was built on one of the islands. In the 1930s, when international air routes were first being extended across long over-water stretches, initially using flying boats, the islands presented a natural solution to the problem of crossing the west Pacific.

Even flying boats, however, cannot safely land in the heavy swells of the open ocean, so before Midway could be used Pan American had to invest large sums in blasting the shallow lagoon clear of submerged coral heads. Once this was

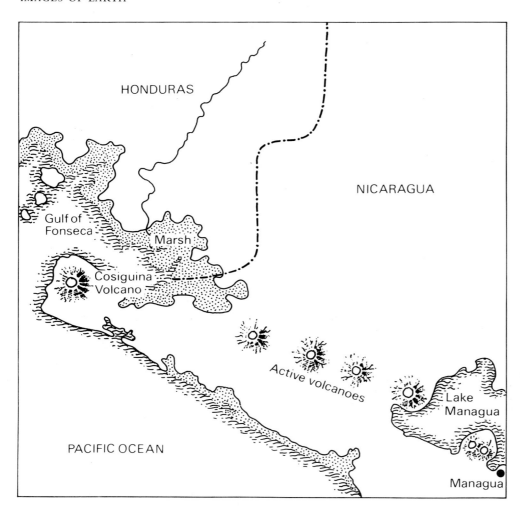

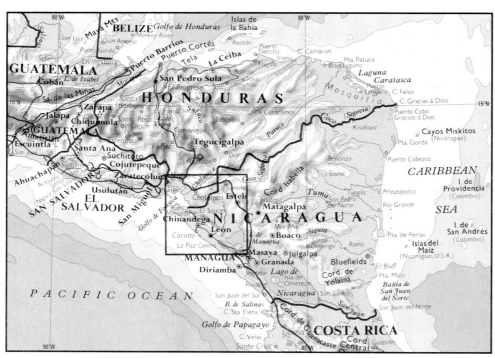

20 One of the most physically and politically unstable parts of the world: Nicaragua, Central America (Landsat, false colour). Image scale 1 cm = 5.5 km; map scale 1 cm = 80 km.

21 The 'unsinkable aircraft carrier': Midway atoll, Pacific Ocean (Shuttle 7, natural colour). Image scale 1 cm = 3 km; map scale 1 cm = 2.5 km.

accomplished, the lagoon provided a safe, protected landing site, and Midway became a rather pleasant stop-over, with a first-class hotel for the first-class passengers. Apart from the delights of a coral atoll, the passengers enjoyed the company of thousands of booby birds, the natural inhabitants of the islands.

When the USA was precipitated into World War II by the Japanese bombing of Pearl Harbour, Hawaii, the tranquillity of Midway atoll abruptly evaporated, and it

found itself in the front line of the fighting. It was soon raided by Japanese aircraft, but escaped serious damage. The Americans rushed to fortify the islands, and constructed both air and submarine bases. Almost overnight, Midway became an 'unsinkable aircraft carrier'. Eventually, almost every part of the tiny amount of dry land was pushed into service, with major airfields filling up both islands, as the image shows.

Although unsinkable, Midway was far from being impregnable. The Japanese were painfully aware that possession of Midway gave the Americans aerial supremacy over much of the west Pacific, and thus put it high on their list of strategic targets. A major battle for the atoll was fought in June 1942. This battle was not only a turning point in the course of the war, but also marked a change in the way that wars were fought. The battle tilted the balance in America's favour, because in their attempt to take the atoll the Japanese lost a large proportion of their front-line carrier strength, and most of their best pilots. It changed the pattern of naval warfare because this was the first major engagement fought almost exclusively by carrier-borne aircraft: the capital ships of both sides never came into contact with each other. With their failure to take Midway, the Japanese lost the strategic initiative in the Pacific, and were forced to cancel invasion plans for other areas. After Midway, they were effectively in retreat.

When the war ended, Midway remained as a pair of airfields surrounded by thousands of kilometres of open ocean. For a while, commercial aircraft continued to make fuelling stops there, but the enormous improvements in aircraft design during the war had given aircraft much greater range, and scheduled services through Midway ceased in 1950.

* * *

When Trotsky and Lenin inspired the Russian communists in their successful revolution in 1917, they hoped that world revolution would soon follow. Within Russia itself, what might have been a genuinely socialist revolution became bogged down in Stalinist misery, with forced collectivization of farms causing millions of deaths from starvation. In China after World War II, once the invading Japanese had been disposed of, Mao Tse-tung was able to ferment a peasant revolution, and in 1949 drove the nationalist Chiang Kai-shek regime into exile in Taiwan.

For a period after World War II, there seemed every reason to believe that China and Russia, the two great Marxist nations of the world, would form a stable alliance. For a while an alliance did exist, and numerous Soviet aid missions were sent to

China to develop its industry and agriculture. In time, however, the ideological differences between the two countries widened, the Russians moving towards a strongly centralized, bureaucratic form of state capitalism, while the Chinese under Mao tried to engender a perpetual peasant struggle, which came to a head with the infamous Cultural Revolution. Russian technicians and advisors were expelled from China, and tensions developed along the exceedingly long frontier, which winds for thousands of kilometres through north-eastern Asia.

The Amur river (Fig. 22) defines the frontier for much of its length. Shown in this image is the area of confluence between the Amur and its important tributary the Zeya. China lies on the south and west edges of the image; the USSR occupies the rest. Much of the Soviet part of the image is taken up by the flat plains of the Zeya-Bureya depression, while the Chinese side is characterized by wooded hills. Notice the enormous size of the fields on the Russian side of the river, and compare them with the tiny fields on Chinese territory, and in the other Chinese images. Many of the Soviet fields are more than 2 km ($1\frac{1}{4}$ miles) in width, a direct reflection of how the social structure and agricultural policy in the two countries differs.

With a vast area of land at its disposal, and a comparatively small population (about 267 million), the USSR has opted for capital-intensive, western-style farming in huge and hugely inefficient farms. The Chinese, also Marxist, but faced with four times as many people (over 1 billion) living in half the land area, have continued their old-style, intensive farming methods based on peasant labour, with even the smallest plots being meticulously cultivated. Little or no mechanical help is available to the Chinese peasants, but in Stalin's Russia the farm tractor was idolized as the key to rural bliss, and millions were produced. It was the use of the tractor that made large fields possible, and which has resulted in the landscape shown on the image.

Although there have been no major conflicts along the frontier, tensions often flare, and the possession of some of the numerous islands in the river is hotly contended. The most significant aspects of the Chinese-Soviet ideological breach are strategic: the Soviet Union is obliged to keep many army divisions along its eastern frontier to secure it against any possible Chinese incursions, and the USA has seized on the breach to ally itself with China, thus strengthening its hand in the Far East.

Although it is the largest country in the

world in terms of its area, much of the USSR is a desolate, arctic land with huge tracts of almost uninhabited tundra and steppe. In contrast, the USA has only about half the land area, but most of it is productive and inhabitable and it supports a population that is only slightly smaller. Although the countries also differ greatly in their approach to State security, one consequence of their physical differences is that it is very much more difficult to hide anything in the USA than the USSR. In fact, the sites of American nuclear missile silos and airforce bases are well known and the subject of popular debate. Such debate is suppressed in the USSR and the huge expanses of the Siberian forests and tundra also make it easy to conceal even the largest missile sites and military bases.

The Defense Department in the USA, of course, has a burning curiosity about such sites, and spends vast sums on satellite observations of them. For the most part, the techniques used and the results obtained by these spy satellites are closely-guarded secrets. Popular gossip has it that the spy cameras are capable of photographing the label on a matchbox, but the effects of turbulence in the Earth's atmosphere make this extremely unlikely. It would be difficult enough to read the label on a matchbox through a telescope at 100 m (330 ft) on a hot day, let alone from 100 km (60 miles). Furthermore, although there are sensors capable of looking through cloud, the resolution that these instruments can achieve is nothing like so good as those of ordinary optical systems. Thus, good clear weather is essential for the best results, and all too often the targets are concealed beneath banks of cloud.

These reservations apart, there is no question that it is possible to learn a great deal from spy cameras. Figure 23 is an example of what can be seen in an image of an unnamed Soviet launch site photographed with the ordinary Hasselblad camera that the astronauts customarily use for their earth photography. This site is at Choybalsan in Mongolia, near the frontier with China. Half-a-dozen separate launch pads can be seen linked together by roads and tracks like nerve cells. There is an airport at the centre, with two runways, a railway yard and many excavated areas. Some of the launch pads are well-established and are surrounded by extensive support facilities; others are apparently new, and incomplete. A number of roads radiate away from the central site into the snowy wastes beyond. While its isolation provides good security in one sense, the concentration of activity in the middle of the empty plains is in itself eye-catching, and the astronauts had no difficulty in locating the site.

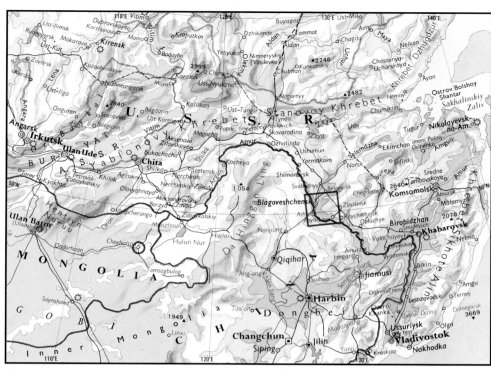

22 LEFT A forgotten frontier: China and the USSR face each other across the Amur river (Landsat, false colour). Image scale 1 cm = 5.5 km.

23 BELOW The Soviet missile launch site at Choybalsan, Mongolia (Shuttle 9, natural colour). Image scale 1 cm = 4.5 km.
The map shows the area of Figure 22 and the location of Figure 23; map scale 1 cm = 200 km.

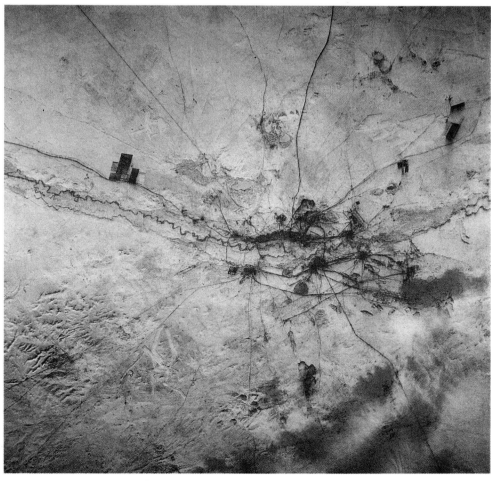

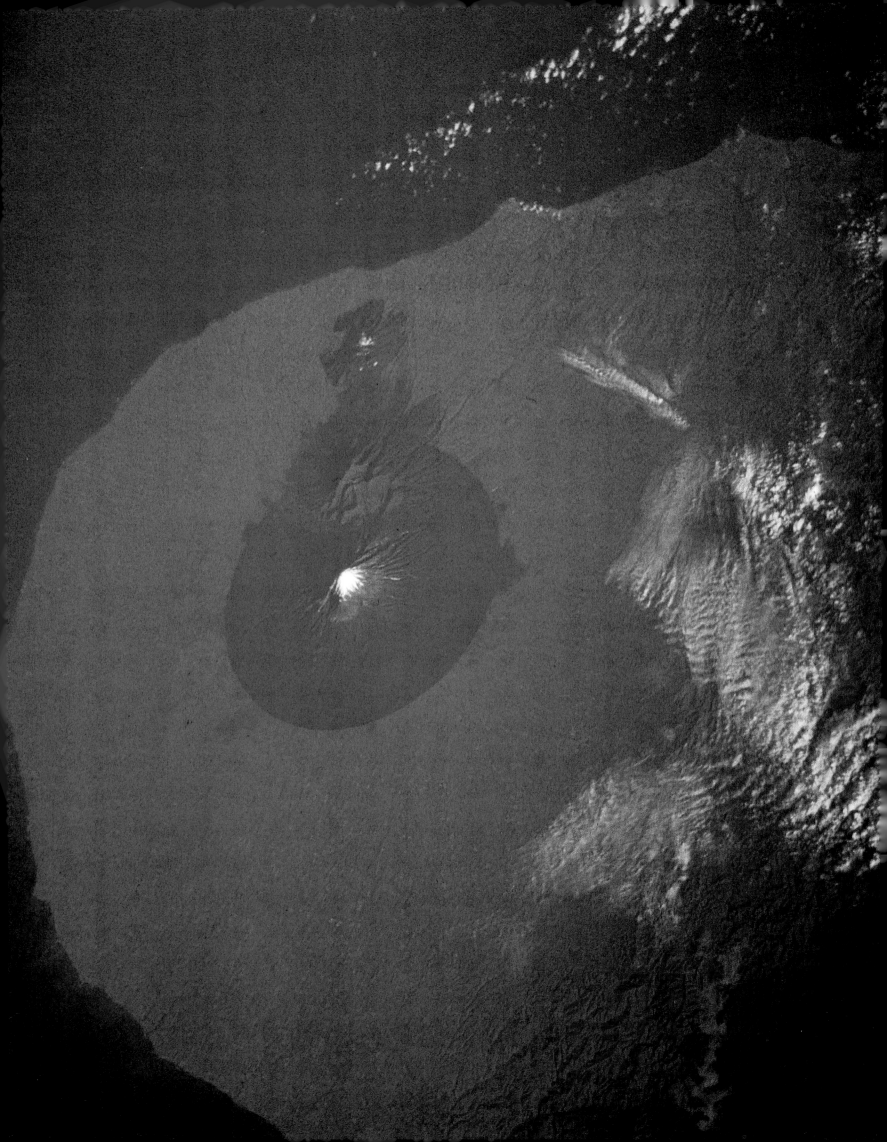

3 Man the Organizer

A voyager arriving by spacecraft from another planet would have to get very close indeed to the Earth before he could see any evidence for the existence of intelligent life. Even using a telescope to look at the Earth from the Moon, our closest neighbour in space, it would be extremely difficult to see anything that might suggest that man existed. At best, it might be possible to see a few dark patches with curiously straight edges where large irrigated areas extend into desert sands. From the altitude at which Landsat orbits (917 km, 573 miles), much more can be seen, and from the Space Shuttle, skimming above the Earth's atmosphere only 280 km (175 miles) up, man-made features are conspicuous even to the naked eye.

The features which are visible have one thing in common: there is something about them which suggests that order has been imposed upon nature; that the landscape has been organized. Agricultural areas are distinctive, especially where fields are large as in the USSR. Rectilinear features such as roads, canals and pipelines also show up clearly, especially in dry areas. Towns and cities tend, unfortunately, to be visible only as grey smudges, concealed by the haze of their own pollution, but the concrete runways of airports and other large features such as dock installations usually show through. Small villages and settlements, generally invisible in themselves, often reveal their locations in the radiating web of roads and tracks that serves them. All these features reveal man the organizer, scurrying busily about his daily concerns, travelling in straight lines from place to place, and imposing his will over forests and deserts, plains and marshland. Sometimes, the effect of this organization is to create, quite unintentionally, patterns that are a delight to the eye.

Mt Egmont, New Zealand (Fig. 24), was photographed by the Shuttle 9 astronauts just after dawn. The 2518-m (8260-ft) snow-capped volcano is casting a long shadow indicating the approximate direction of the midsummer sunrise. The perfect symmetry of the volcanic cone is echoed by the sharply defined green ring encircling it. Beyond the dark green circlet, fields and pasture-lands extend in a lighter green mosaic to the coast. Although the result is remarkably pleasing, there was no aesthetic intent in the minds of those responsible for the dark green boundary. It simply marks the edge of a national park whose limits were prosaically defined by laying out a circle, centred on the volcano summit, with a radius of 10 km (6 miles). Within the park, dense forests clothe the slopes of the volcano as far as the high meadows below the snow-line; outside the park boundary, the original forest has been entirely cleared for farming. The volcano was first sighted by Captain Cook in 1770, who saw it in its primeval state and who would probably be amazed if he could see the changes that two hundred years have wrought. The park is immensely popular with tourists and skiers. In the image the snow cover on the volcano is very thin, and far from ideal for skiing, but this is because the image was acquired during the height of the southern summer. For the same reason, the volcano's shadow is extended to the northwest, rather than the west.

On Mt Egmont, the hand of man has drawn quite lightly, although the results are eye-catching enough. Detroit, USA, represents the other extreme, where urbanization and heavy industry have sprawled over a huge area (Fig. 25). First settled by Antoine de la Mothe Cadillac in 1701, and situated at a strategic site commanding the St Lawrence Seaway, it is easy to see why a city grew up at Detroit. Why it became the focus of the world's automobile manufacturing industry is less obvious. It is conveniently located for the supply of raw materials – iron ore from the banded iron formations on the shores of Lake Superior, coal from Pittsburgh – but a more important factor may have been that carriage manufacturing was well-established in the city when Henry Ford began building the horseless variety there in the early 1900s.

The most conspicuous element in the image is the complex of freeways that run through the city, and the bright network of runways and taxiways that is Detroit's main airport, the Wayne County airport. A

24 Mt Egmont, New Zealand, with the dawn sun casting a long shadow over the circle of the national park (Shuttle 9, oblique view, natural colour). Image scale 1 cm = 2.5 km; map scale 1 cm = 25 km.

25 OVERLEAF Detroit, Michigan, USA, centre of the world's automobile industry (Landsat, Thematic Mapper, false colour). Image scale 1 cm = 2.6 km; map scale 1 cm = 25 km.

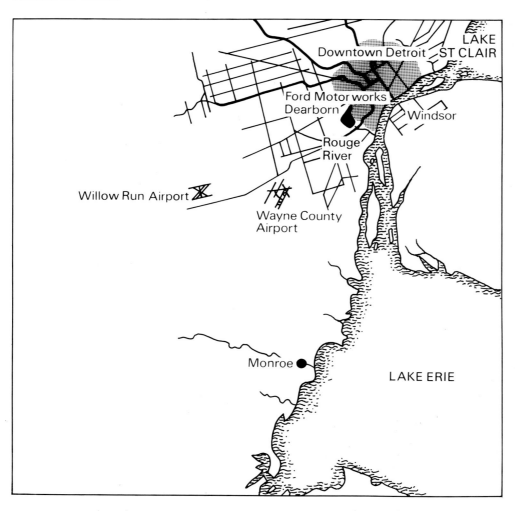

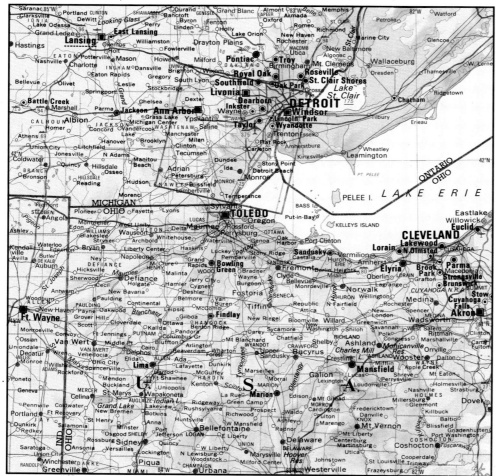

little further west of this is the somewhat smaller Willow Run airport, its runways intersecting to form a pattern like a Union Jack. The large body of water in the lower right is Lake Erie, and that in the upper right is Lake St Clair, a curl of sediment-laden water showing clearly where the channel connecting the two lakes enters Lake Erie. Deep, clear water shows up black in false colour images, since water absorbs infra-red radiation. Only very shallow, salty or sediment-laden waters show up in other tones. Despite the lack of obvious heavy sediment, Lake Erie is one of the most polluted lakes in the world, the consequence of accepting decades of industrial effluent and the excess fertilizers that are leached from cultivated land. The chequered pattern of fields in the western part of the image is an indication that the industrial city is surrounded by prime farming country.

While Henry Ford was responsible for establishing the automobile industry in Detroit, other manufacturers, including the rival giant, General Motors, later arrived to take advantage of the skills developed by city workers. The Canadian city of Windsor which grew up across the river from Detroit has also developed an automobile industry. The two cities are linked by bridges, tunnels and ferries in what must be one of the busiest international frontiers in the world. It is slightly surprising in view of all the hectic manufacturing industry that there is little direct evidence for it on the image. The fact that there is no obvious pall of fumes, or belching smokestacks, is the result both of

favourable winds, which sweep fumes away towards Canada, and efficient smoke control regulations. So large is the scale of industry, however, that it reveals itself in other ways. Ford's main car plant in the suburb of Dearborn, for example, covers nearly 4 sq km (1.5 sq miles) and shows up as a distinctive dark polygon at the top of the black prong of the Rouge river (see sketch map).

A Shuttle astronaut orbiting over the central USA has little need of a map to tell him which way is north. North America is all laid out on a north-south/east-west grid system, so all field boundaries run either north-south or east-west. Major roads and railways follow the same pattern all over the flat plains and prairie regions. Detroit, however, was laid out originally on a northwest-southeast grid, to fit in with the natural trend of the Detroit river. As the settlement grew from a village in the 1700s to a major city in the 1900s, it spread outwards, preserving the northwest grid as it grew. Eventually, however, it intersected the north-south grid of the rural areas, and an uncomfortable suture between the two grids has resulted, visible in the dark belt surrounding much of the inner city.

The city has now spread far beyond the intersection of the two grids. In northwest Detroit the progressive conversion of farmland into suburbs is especially clear,

26 The real America: Topeka and Kansas City, Kansas, USA (Landsat, Thematic Mapper, false colour). Image scale 1 cm = 5 km; map scale 1 cm = 25 km.

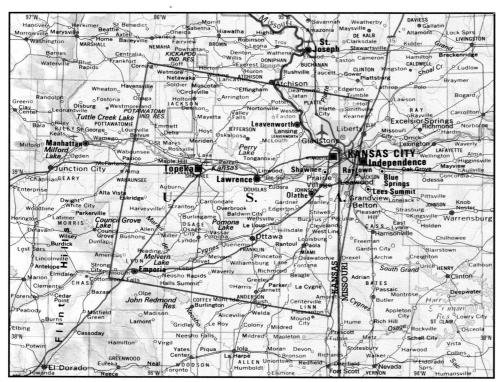

27 Milton Keynes, Buckinghamshire, a
new city spreading over the English
landscape (Landsat, Thematic Mapper,
false colour). Image scale 1 cm = 1.5
km; map scale 1 cm = 12.5 km. Image
reproduced courtesy of Milton Keynes
Development Corporation.

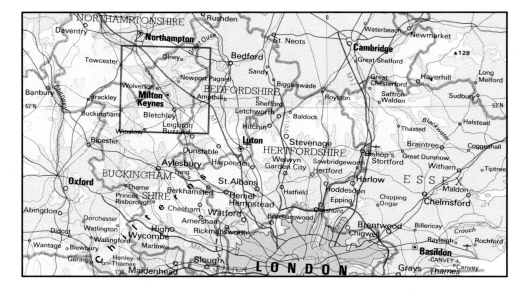

with the city organized into freeways, major highways and lesser local roads that divide the countryside up into a hierarchy of smaller and smaller squares. The most prominent squares are exactly one mile (1.6 km) across: scarcely the most imaginative form of town planning.

The same one-mile square north-south grid pattern covers most of the heartland of the USA. It can be seen clearly again on Figure 26 which shows part of the Great Plains and includes Kansas City, almost exactly in the centre of the country. The flat, open nature of the plains made them ideally suitable for the development of large farms, and some of the biggest fields visible here are one mile square. These vast areas also lend themselves particularly well to mechanized farming and it is not surprising that this is the home of the combine harvester. The Great Plains are enormously important to the economy of the USA and the world as the endless chequerboard of fields provides a large proportion of the Western World's grain. Kansas has traditionally been the largest wheat-producing state in the USA, but it is not solely a wheat producer – Kansas City, for example, is second only to Chicago in the size of its stockyards and meat-packing plants.

On the right of the image is the bluish-coloured urban sprawl of Kansas City. There are actually two 'Kansas Cities', since the metropolitan area straddles the state line between Kansas and Missouri, which follows the meandering course of the Missouri river. Thus, there is often a confusion between Kansas City (Kansas) and Kansas City (Missouri), since each is a distinct administrative unit. Perhaps reflecting the uniform, even monotonous patterns of the Great Plains, and its isolation from the more cosmopolitan coastal cities of the USA, Kansas City (Kansas) and the state generally are traditionally strongly conservative, Republican strongholds. Many other mid-western cities in the same geographic setting have similar political alignments. Kansas was the first state to introduce prohibition and even today restrictive licensing laws limit the sale of beers and wines. Across the river in Missouri, state liquor laws are more liberal, although in other respects the state is almost as conservative as Kansas. Kansas City (Missouri) has much outgrown its Kansas counterpart and now has a population that is approximately three times as large. It has been seriously suggested that the liquor laws are partly responsible for this, the more liberal Missouri city attracting more outside investment and development.

The present metropolitan complex grew from small beginnings, when white settlers purchased land originally settled by the Wyandot Indians at the confluence of the Missouri and Kansas rivers. In 1866, railways reached the town and it grew rapidly in size, as did others along the Kansas river such as Lawrence (centre) and Topeka (left centre). Apart from these cities, there is little to interrupt the chequerboard pattern of fields, except for the river valleys and a few reservoirs. The valley of the Missouri is particularly striking, consisting of large blocks of solid red, with fewer obvious field boundaries. This strip along the river defines the floodplain, which is wetter than its surroundings and unsuitable for growing grain. It is ideal for pasture, however, and the red shows that the fields are almost all covered with green grass, contrasting sharply with the blues and greys of the fields over much of the rest of the image, many of which have been freshly ploughed. Intermediate red and brown colours represent crops of varying degrees of ripeness. The fields between Kansas City and Lawrence also conceal a more sinister crop: neatly laid out rows of Minuteman missile silos.

Although it is most expansively developed in the great plains and prairies of the USA and Canada, there is nothing peculiarly North American about the regular north-south grid pattern. Roman cities were very strictly laid out on a square pattern, and the concept is one that has survived through the ages and spread to many lands the Romans had not even dreamed of – even the smallest South American cities are laid out in a regular grid, centred on the inevitable Plaza de Armas. Although the Romans introduced the grid pattern into many of the cities of Europe, it has in most cases been superseded by a disorganized but charmingly irregular maze of streets and lanes intersecting at improbable and inconvenient angles. Though not nearly so simple for traffic planning as grid cities, and thus frequently the sites of monumental traffic jams, these apparently unplanned cities are generally thought to have more appeal than grid cities, both by those who live in them and by the tourists who visit them.

Given this preference, it might seem odd that, when planners have the chance to create new cities, they often choose the old grid pattern. Milton Keynes, spreading like a cancer on the tranquil green fields of rural Buckinghamshire, provides an excellent example (Fig. 27). The city was conceived in the 1960s with the best of intentions. It was planned to ease population pressure in London, 80 km (50 miles) to the south, and to provide new jobs and pleasant living conditions for thousands of people in a completely new, highly-organized environment. Construction began at the end of the sixties, and the city now stretches from Bletchley to Wolverton, incorporating both these old towns.

If the new city has proved a disappointment to its new inhabitants, it is not for want of benevolent planning. Unfortunately, however, the master plan for the city was conceived at a time of economic buoyancy, when energy was inexpensive and looked as though it would become even cheaper. Thus Milton Keynes was planned on the assumption that each family would have at least one, or probably two cars, and was laid out on the basis of a very low population density, with residential, shopping and industrial areas united by a grid of major roads (divided highways). Shortly after the plan had crystallized to the point of irrevocability, however, the energy crisis blighted hopes that life in Milton Keynes would be car-oriented and, more seriously, the economic recession at the end of the seventies meant that, rather than having two cars, some families had none at all. But human beings are endlessly adaptable and the city survives and continues to grow. Hundreds of trees have been planted to give a rural ambience. Those unfortunate enough to be without a car in this brave new city may curse the planners as they trudge from residential area to office or bus stop, but they certainly see plenty of trees.

City planners do not have many alternative strategies open to them. Apart from the square grid, the only other geometrical possibility is based on radial and circumferential roads, so that the city map ends up looking like a spider's web. Moscow (Moskva) illustrates this arrangement perfectly and appropriately: the Kremlin sits at the centre of the web. In a curious way, the highly centralized geography of the city (Fig. 28) mirrors the extraordinarily centralized Soviet government and administrative structure: the lines of command extend radially from the Kremlin to the most distant parts of the country.

The core of the city, its oldest part, is centred around the Kremlin on the steep northern bank of the Moskva river where the first fortifications were built in the twelfth century. Throughout the early history of Moscow, there was heavy emphasis on defence and the growing city was encircled with walls. These now define the course of the present inner ring roads – the Garden Ring, for example, follows the course of the sixteenth-century city limits, which were defended by earthworks, wooden walls and a deep moat.

Modern Moscow is vastly more extensive than the walled city. Much of it has developed in the forty years since World War II. Enormous efforts have been expended to make Moscow a show place, the pride of the USSR, a city that the Russians can use to demonstrate to themselves and to the West that socialism has really worked. Vast areas of decrepit buildings were demolished after World War II, and huge new high-rise complexes constructed in the outer suburbs. Some of these enormous building complexes are visible in the image on the south side of the city – the individual blocks are so large that they are catching the sun and casting long shadows.

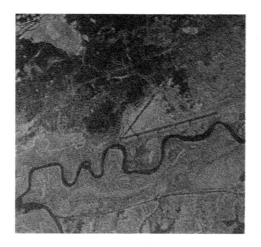

No fewer than five large airports are situated around Moscow, illustrating the volume of traffic that the city generates. Whereas most major airports in the USA can take international traffic, Moscow's Sheremetyevo and Domodedovo airports handle the vast majority of the USSR's international flights. One of the most interesting airports is not a civil airport at all, but a secret one at Ramenskoye, 32 km (20 miles) to the east at the edge of the city, on the north bank of the Moskva river (see enlargement). With a runway over 5500 m (18,000 ft) long, Ramenskoye is one of the largest airfields in the world. Such a long runway is not needed for any ordinary aircraft, and in fact it is used for test flights of the USSR's answer to the Space Shuttle, carried piggy-back on a large aircraft. The Shuttle itself went through similar trials during its development. It says much for the quality of US monitoring of Soviet facilities that they were able to observe the space vehicle and its carrier after an embarrassing accident, when the carrier aircraft ran off the runway, and both craft had to be salvaged with heavy cranes and lifting gear.

In images of developed countries, it is the road networks that tell us most about the efforts of man. In other areas, and especially in China, it is the waterways. Figure 29 covers part of the Hopeh (Hebei) province, one of the most densely-populated areas in the world. Hundreds of small villages dot the northern part of the image (visible as light brown spots against the red background), and the city of Tientsin (Tianjin), third largest in China, forms the grey smudge at the head of the dark, meandering Hai He river. The outer suburbs of Peking (Beijing) can be seen at the top left of the image, connected to Tientsin by a narrow thread of road. This is perhaps the most important road for motor traffic in all China, and is one of the few that is easily discernible on Landsat imagery. By contrast, many major canals and canalized rivers are easily spotted. For example, the Grand Canal, which originates over 1600 km (1000 miles) to the

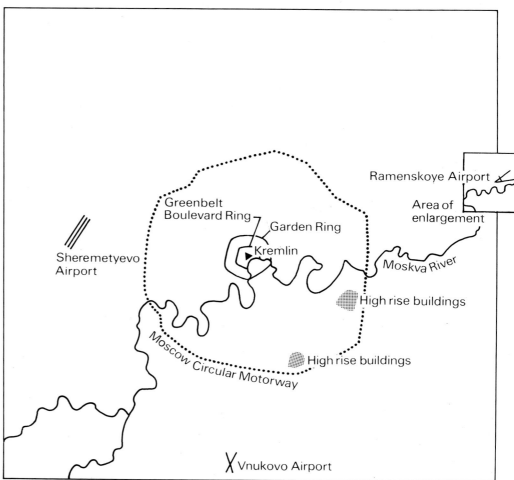

south, can be seen crossing the centre of the image and leading into Tientsin, where it terminates.

Like many other parts of China, the area around Tientsin is extremely low-lying – the city itself is 4.5 m (15 ft) or less above sea-level and flooding is a perpetual scourge. One reason why the waterways are so conspicuous on the image is that they have been confined by massive artificial banks, or levees, as part of river management projects. These are best seen in the waterways at top right where the levees are visible as extremely narrow grey bands paralleling the blue of the water. Most of the major channels in the southern

part of the image are also leveed, although the levees themselves are harder to make out. These levees were largely constructed by communal labour teams without the aid of machines – a feat that almost defies imagination.

Along the coast, the pale grey band is a zone of intertidal mudflats; these are more than 10 km (6 miles) broad at low tide. The large, dark blue enclosures just behind the mudflats are artificial salt-pans, used for obtaining salt from sea-water by evaporation. Different stages in the process of extraction can be seen in the varied tones, the paler blue representing the most concentrated salt. Inland from and slightly

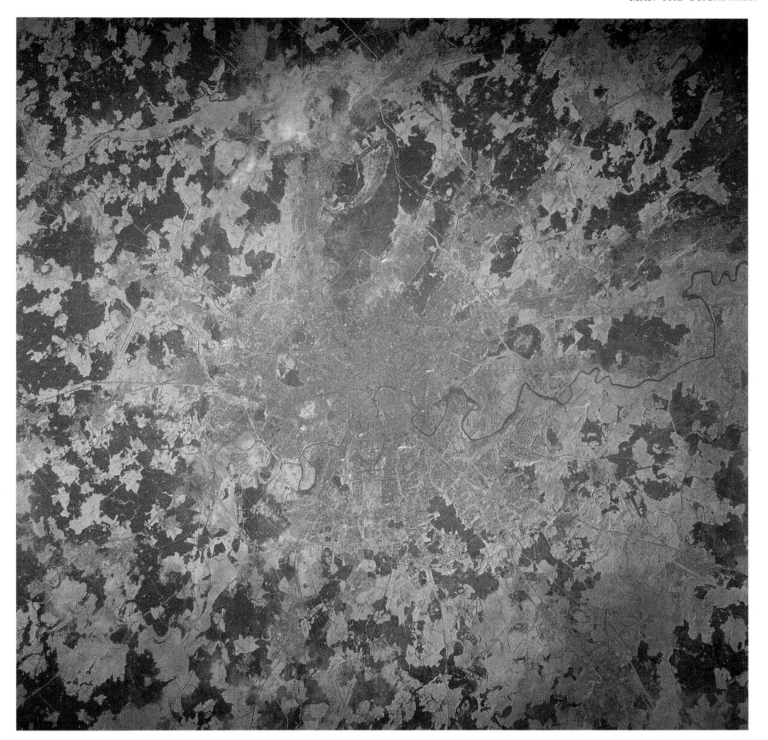

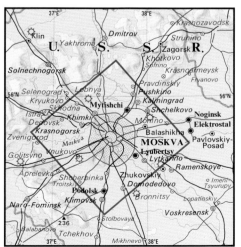

28 The spider-web plan of Moscow, USSR, mirrors the highly-centralized Soviet administration (Shuttle 9, natural colour winter scene). Image scale 1 cm = 3.7 km; map scale 1 cm = 25 km. The enlargement (above left) shows Ramenskoye airport to the southeast of the city, used for test flights of the USSR's answer to the Space Shuttle. Scale of enlargement 1 cm = 2 km.

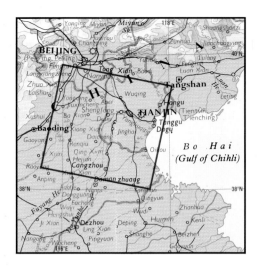

29 Tientsin and the waterways of Hopeh Province, China, one of the most densely-populated regions in the world (Landsat, false colour). Image scale 1 cm = 9 km; map scale 1 cm = 60 km.

below the salt-pans are dark red enclosures; these are vast rice paddies, irrigated by the canals that can be seen. The black area in the middle of the paddy is open water. There is a second large area of paddies on the left of the image.

These paddies, large as they are, are insignificant compared with those around the Dongting swamp, the rice bowl of China (Fig. 30). Vast areas of continuous rice paddies show up in bright red tones in the top right of the image, evidence of the extraordinarily intensive agriculture. Smaller belts and patches of rice paddies are strung out along the rivers in the upper left. Because of China's pressing food shortages, scarcely a square metre of fertile ground is left unused. The low-lying area of the Dongting, with its many rivers and lakes, provides almost ideal conditions for rice cultivation. The growing period lasts eight to eleven months, and two or sometimes three crops can be raised. Not surprisingly, the rural population density is extremely high, though there is little direct evidence of this on the image, apart from the rice paddies and some canals. At lower right, the major city of Changsha on the right bank of the Xiang river is just visible. The city, founded in the third century BC, is a port and a major industrial centre, serving the rural communities in the hinterland.

Charting the innumerable small, shifting rivers and lakes in the area was a perennial problem for Chinese cartographers, but now satellite images provide very accurate detail. Three rivers drain into the Dongting from the south; the Yuan, Zi and Xiang. At the top right of the image, a few large meanders of the Yangtze (Chang) can be seen, and it is this river, the longest in China, that is probably most responsible for the swamp. Rising high on the Tibetan plateau, the Yangtze winds for no less than 5500 km (3440 miles) before entering the sea near Shanghai. Just west of the image, the river runs through a mountain range in a major series of narrow gorges, but once free of the mountains it seems to lose its way temporarily in the vast inland basin that is the Dongting swamp.

In the geological past, the main channel of the Yangtze probably came down further to the south, where the lake is now, but as more and more sediment accumulated in the lake basin, the river channel shifted northwards. At present, the lake forms a kind of regulator for the flow of the lower Yangtze, ponding up seasonal floodwaters. Eventually, however, sedimentation will fill the entire basin, providing more land for rice.

Although the area of the USSR is enormous, only a small proportion of it is good agricultural land. Most farming is concentrated in a wheat-belt that extends from the Ukraine in European Russia as far as Siberia in the east. Figure 31 illustrates the heart of the wheat-belt, and is centred on the city of Troitsk (lower centre); the capital of the region is Chelyabinsk, a major industrial centre, showing at the top of the image, just left of centre. The prominent river valley is the Uy, which ultimately reaches the frigid Arctic Ocean in the far north of the Soviet Union. Myriads of shallow rounded lakes in the north and hundreds of square fields in the south present an odd dichotomy in this image, the one natural, the other artificial. The lakes, which range from more than a kilometre across to small pools, were probably formed by wind erosion of sediments. Much of the USSR's great steppes (roughly the equivalent of the American prairies) are underlain by fine-grained wind-deposited sediment known as loess, laid down when the climate was cold and extremely dry during the last Ice Age. Loess is easily sculpted and hollowed out by the wind and, as the climate warmed up and became wetter at the end of the Ice Age, the wind-blown hollows filled up with water. Some of them have since begun to dry up and are now ringed with white salt, or, like the one near the centre of the image, are almost entirely covered with a salt crust.

Salt-encrusted lakes show that the area is still a dry one, especially in summer when this image was acquired. The climate is suitable for wheat, however, as the great expanse of enormous collective farms demonstrates – many of the fields are over 2 km (1.25 miles) across. Almost all the fields in the southern part of the image have bluey-grey tones, rather than the bright reds indicative of actively growing green vegetation. This suggests that the fields are lying fallow, with bare earth exposed, possibly after a harvest of winter wheat, or before spring wheat has come into leaf.

Inspection of a single image obviously cannot tell us why the fields were bare, but the advantage of satellite imagery is that it can be used to obtain repeat coverage, so that a close eye can be kept on what is growing. The Americans have gone to great lengths to develop techniques for monitoring Soviet agriculture, and Defense Department experts are now able not only to tell what kinds of crops have been planted, and when they will be harvested, but also what the harvest is likely to yield. Their forecasts have in some cases been better than those of the Soviets themselves. Desperately inefficient collective farming methods coupled with unfavourable weather have resulted in such poor harvests in recent years that the USSR has had to buy large quantities of grain from the USA in order to feed its people and cattle. Although the Soviets have driven some sharp political bargains in the grain deals, the Americans' detailed knowledge of the Soviets' likely grain shortages gives them a powerful card in the international poker game.

Fields of similar size to those in the USSR are also used in Australia, but for quite different reasons. Although a considerable quantity of wheat is grown, most of Australia's agricultural land is used for grazing sheep or cattle. Because the climate is so dry over so much of the continent, enormously large areas are needed to support grazing flocks. In the driest parts, a single sheep may need as much as 8–12 hectares (20–30 acres) in order to find enough food for itself; a flock, therefore, may be dispersed over hundreds or thousands of hectares. Many individual sheep stations are over 400 sq km (100,000 acres) in size, and some are as large as 4000 sq km (1,000,000 acres). Western Australia has many such large ranches, and Figure 32 illustrates some of them near the Yarra Yarra lakes, the white, dried-up salt lakes at lower left. Marching boldly across the image is the step-like margin between natural scrub (right, dark) and the areas which have been painstakingly cleared and divided into fields for agriculture. The native scrubland is so barren and poor that it has to be artificially fertilized with phosphates to provide grazing of any value. Over 4000 sq km a year are being brought into agricultural production with the help of government grants for clearing and fertilizing, and as a result the margin marches further inland every year.

As in the Soviet image, the dried-up lakes speak of an arid climate, but the difference is that the Yarra Yarra area can be much hotter, with summer temperatures above 40°C (100°F). Coupled with the low rainfall, this not only means that the grazing is poor but also that there is a high risk from grass and scrub fires, which are a curse of Australian life. Since agriculture is necessarily thinly spread over this barren land, so too is the population. A total of only about 1,300,000 people live in the whole of Western Australia – an area eight times that of the British Isles – and by far the majority live in the single large city, Perth. So the remainder are very thinly spread indeed, in communities such as Mullewa, strung out along the line of the road and railway that cross the north part of the image and which lead to the small port of Geraldton on the Indian Ocean coast. Although the grazing lands can be extended by further encroachment into the scrub country, it will never be possible for Western Australia to be thickly populated

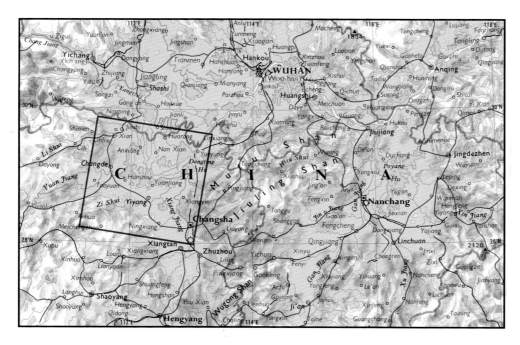

30 Dongting swamp, rice bowl of China (Landsat, false colour). Image scale 1 cm = 5.5 km; map scale 1 cm = 60 km.

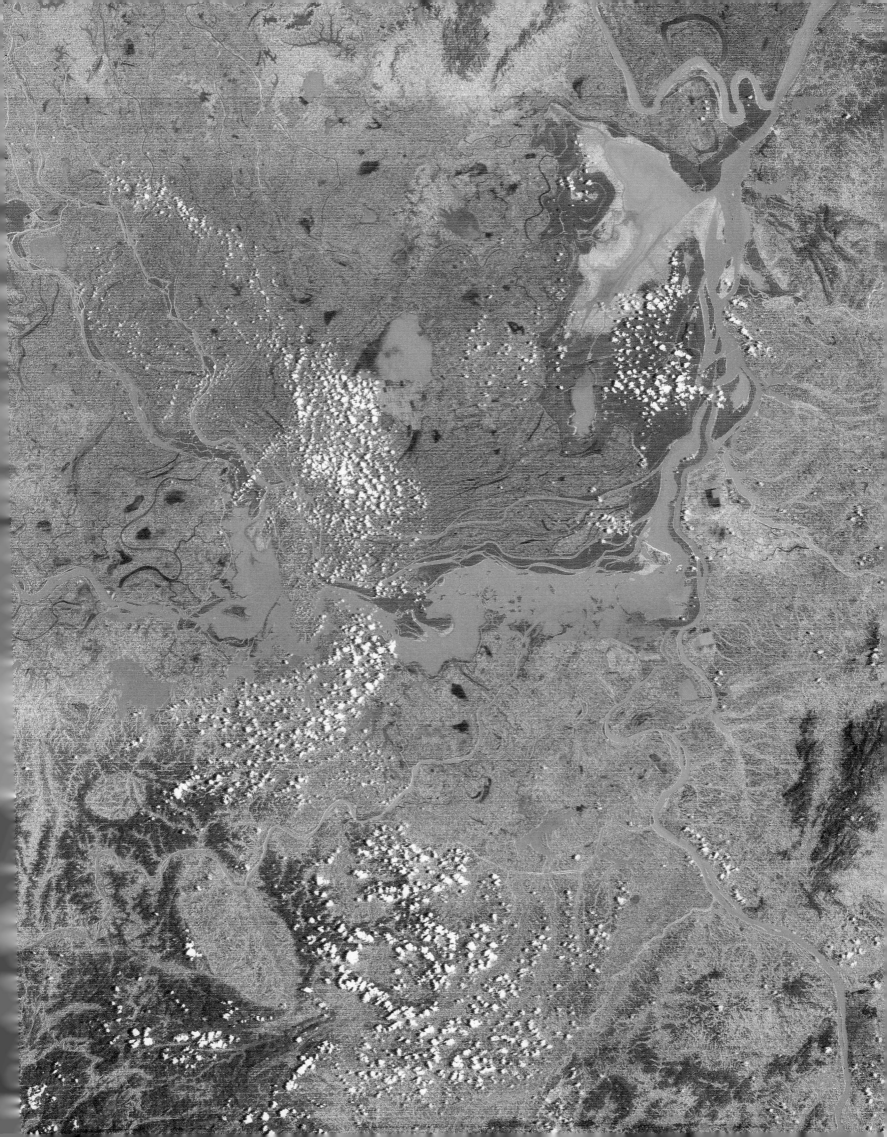

or intensively farmed, since it is simply too dry. Most of the state has less than 25 cm (10 in) of rain a year. There are no permanent rivers of consequence in the whole vast area, and although underground artesian water can be pumped for irrigation, its supply is not infinite.

Throughout history, water or the lack of it has exerted a powerful influence on what man can or cannot do with his environment. Nowhere is this dependence on water more clear than in the Nile Valley (Fig. 33). For millennia the great civilizations living along the Nile depended on it entirely for their water supplies, and for the fertilizing silts that its annual floods brought with them. Two distinct agricultural areas developed: a narrow, irrigated belt following the course of the Nile itself, and a broad fan in the delta area which the Nile has built up through the course of geological time by depositing its burden of fertile silt into the Mediterranean. Just downstream from Cairo (lower left in the image) the great river divides into two channels, the Rosetta and the Damietta, and the sinuous course of the latter can just be seen at top left. From the earliest days of prehistory, through the glorious pharaonic dynasties up to the twentieth century, daily life in Egypt was dictated by the flow of the Nile. But the construction of the Aswan High Dam in 1970 changed traditional ways of life. Designed to increase the volume of water available for irrigation, the dam has achieved its principal aim, but has also had a number of much less desirable effects. For example, the lower reaches of the Nile no longer receive their annual increment of fertile soil, and construction of numerous canals has

encouraged the spread of the water-borne disease bilharzia.

Underlying the decision to build the Aswan Dam was Egypt's pressing need to feed its exploding population. The image provides a fascinating insight into another aspect of this problem. A well-established belt of irrigated farmland follows the course of the Ismailiya Canal, joining the Nile with the Suez Canal, and passing through the improbably named town of Zagazig. Both north and south of the canal, clusters of small circles can be seen drawn on the sandy desert surface. These circles are quite new; photographs taken of the same area by Gemini and Skylab astronauts only a decade ago show only empty, unbroken, sandy desert. The circles are examples of a new agricultural technique, known as centre pivot irrigation. A long, mechanized arm, carrying sprinklers, trundles slowly around a central axis, supplying a controlled stream of water to the circle it describes. The arms can be extremely long, hundreds of metres, and the discs that they trace out make a startling contrast to the normal grid patterns of traditional agriculture when seen for the first time from space. In this picture, only a small proportion of the circles are green with crops. Most are still dusky and sandy, but they show that remarkable strides have been made in bringing more land under cultivation by using new and efficient techniques.

31 Troitsk and Chelyabinsk, USSR, the heart of the Soviet wheat-belt (Landsat, false colour). Image scale 1 cm = 5.5 km; map scale 1 cm = 50 km.

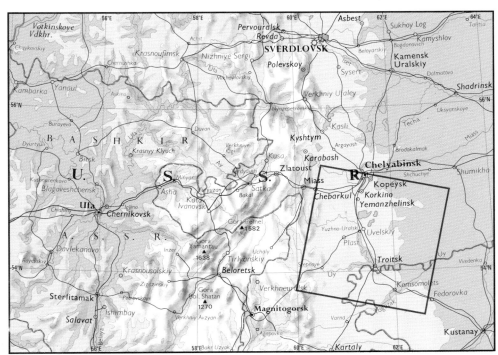

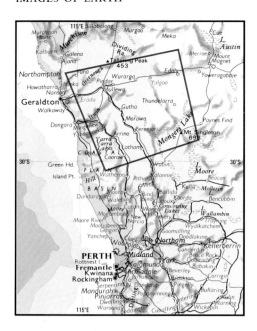

Apart from the new centre pivot irrigation scheme, this image also shows the hand of man clearly in the Suez Canal, still an impressive piece of engineering after more than a hundred years. Port Said (Bûr Sa'îd) is at the northern end of the canal; Suez (El Suweis) is at the southern, on the Red Sea. For over a hundred years the canal was of enormous strategic importance, and was the cause of one of Britain's last imperialist exploits in 1956, when a military task-force was parachuted into the canal zone in response to Egypt's unilateral decision to nationalize the canal. Oil was the chief reason for its strategic impor-tance, since the shortest route to Europe for tankers from the Gulf ports lay through the canal. Now, however, its importance has declined somewhat, since crude-oil tankers have grown so large over the last decade that they can no longer get through the shallow canal. Near the centre of the picture, the Great Bitter Lake was so shallow that a channel had to be cut in the sands of the lake bottom. Constant dredging is needed to keep the canal open even to vessels of modest size, while the great supertankers are forced to steam in a steady procession from the Gulf ports around the Cape of Good Hope to Europe.

32 The hand of man is unmistakable in the Yarra Yarra Lakes area of Western Australia (Shuttle 5, natural colour). Image scale 1 cm = 10 km; map scale 1 cm = 80 km.

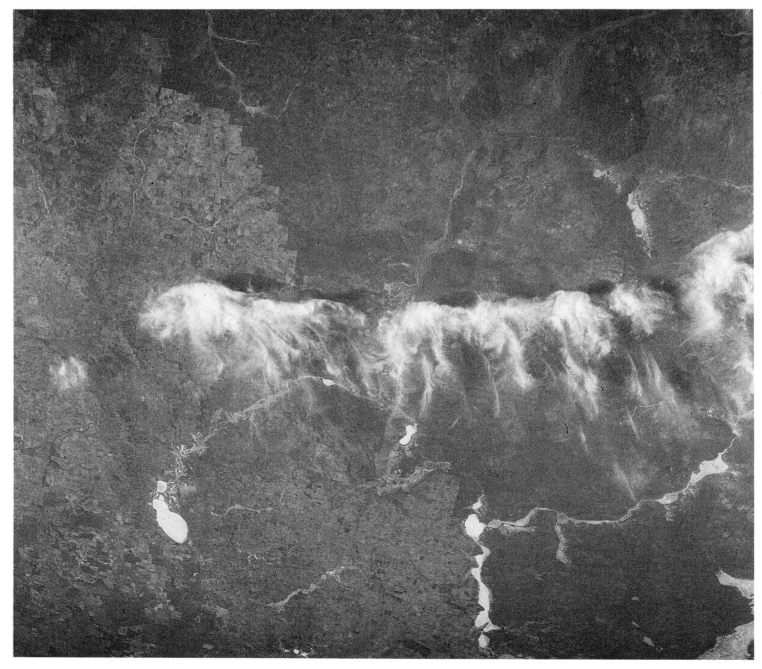

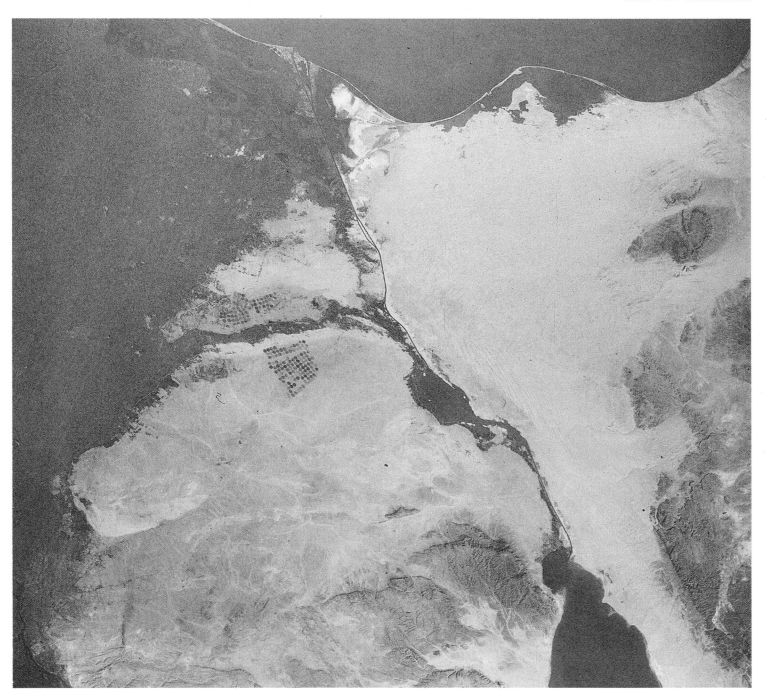

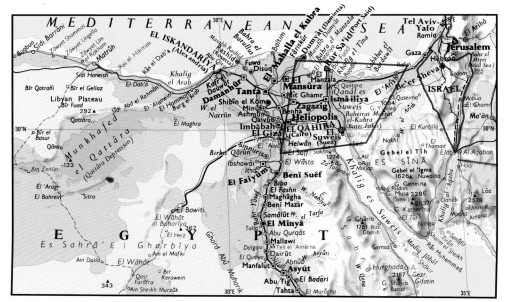

33 New centre-pivot irrigation circles stud the desert beyond the traditional fertile areas of the Nile delta (Shuttle 9, natural colour). Image scale 1 cm = 10 km; map scale 1 cm = 80 km.

4 Earth Under Pressure

When any change is made in our environment, such as the construction of a new road, or the felling of a forest, it is often said that the first generation affected resents the change, the second accepts it, and the third regards it as part of the normal, everyday background. In this way, profound changes have taken place on the face of the Earth which a century ago would have been unthinkable, but which today are accepted unhesitatingly. Who would have thought fifty years ago that the quiet fields around the English village of Heathrow would today become the site of the busiest international airport in the world, embedded in a metropolitan complex that extends over 50 km (30 miles) along the River Thames?

Each individual change, of course, is quite small. When Heathrow opened, the amount of traffic it handled was tiny, and Londoners used to come out on sunny afternoons for the pleasure of seeing aircraft taking off and landing. Now, however, after decades of gradual growth, they are more likely to shake their fists at the big jets climbing noisily out over their suburban homes, drowning conversation and causing ghostly, wavering images on countless televisions.

Apart from the slow rate at which such changes take place, another reason why they are so universally accepted is that each one of us is familiar with only a small area. The much larger perspective that one gets from space helps us to see such developments for what they really are, and brings home the magnitude of the changes that have been wrought on the Earth over the last few hundred years, and, most seriously, in the last few decades.

At first glance, the huge areas of farmland that are so impressive on Landsat or Shuttle imagery might seem harmless enough, sometimes even attractive, their tessellated patterns pleasing to the eye and the mind. But the fields replace natural forests or grasslands, and the intensive clearing, fertilizing, soil sterilization, pest eradication and genetic selection processes

that are essential to modern farming have caused vast mortality to plant and animal species. There is a sinister significance to the orderly pattern of wheat fields in images of the American prairies – here a rich variety of life has been replaced by what amounts to a single species of grass. This is the price that has to be paid, if the Earth is to continue to sustain the steady increase in the numbers of a single animal species, man. Monoculture of selectively-bred crops on this scale also brings its own risks. Encouraging a single strain to the exclusion of related varieties has inevitably diminished the available gene pool. This could have serious implications should disease wipe out the highly-cultivated strain as there might not be alternatives available.

Figures 34 and 35 illustrate one aspect of pressure to increase agricultural production: the need to farm the most barren, unpromising lands. Australia is much the driest and flattest of the continents. Most of its interior is a vast inland basin, unrelieved by anything other than low, scrubby hills and dry, dusty river- and lake-beds. Although a continent in size, the whole of Australia contains fewer people (about 15 million) than many metropolitan complexes, such as New York/Boston or the London area of Britain. To fly over it in an aircraft is to contemplate a seared land of sombre reds and tawny browns, where the eye seeks anxiously for the reassuring sight of the hand of man, but where this is often lacking for hundreds of monotonous kilometres at a stretch. From space, Australia gives the same impression of a time-worn, withered land, but with the broader perspective that the greater altitude brings, many subtle relationships between man and his environment become clearer.

In Australia, the difference between regions with some agricultural potential and those with none can be measured in terms of only a few centimetres of annual rainfall. Only the northern and eastern coasts of the continent receive bountiful rain with annual totals of 75 cm (30 in) or more; most of the centre receives less than 25 cm (10 in). In the intermediate zone between coast and interior, agricultural life is permanently at the mercy of the

34 A filigree whorl of dry gullies: the Diamantina river, Queensland, Australia (Landsat, false colour). Image scale 1 cm = 5.5 km. See p. 78 for location map.

N

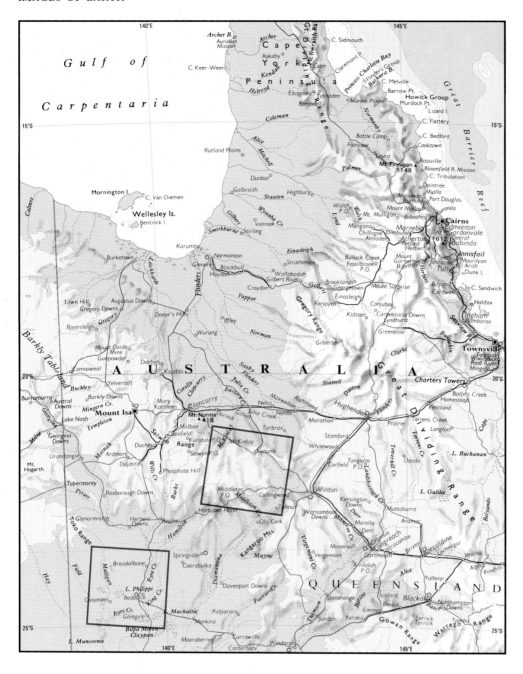

The map shows the area of Figure 34
(above) and Figure 35 (below). Map
scale 1 cm = 80 km.

35 The desert's edge: sand-dunes of the
Simpson Desert march across a vegetated
valley, western Queensland, Australia
(Landsat, false colour). Image scale 1 cm
= 5.5 km.

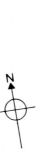

rainfall, years of terrifying drought often being followed by sudden, equally damaging floods. Figure 34 shows the great loop of the Diamantina river in the state of Queensland, as it sweeps round from its watershed on the Selwyn Hills and drains southwards towards Lake Eyre, a great inland basin. The hills are low, reaching only 300 m (980 ft), but the effect they have had on the development of the drainage pattern is striking. Notice the dendritic (tree-like) patterns of the gullies on the inside of the loop. These suggest relatively rapid erosion, with the heads of the gullies nibbling away the intervening ridges to the point where some opposing gullies appear to be about to coalesce. This pattern of erosion may result from overgrazing of what little vegetation survives on the hills in this area of marginal rainfall.

Despite its arid, barren landscape, Queensland is world-famous for its production of beef and wool, and has many of the largest farms or ranches on earth. Called 'spreads' or 'stations', over 2000 farms greater than 400 sq km (154 sq miles) exist in western Queensland, often with not more than 80 animals per square kilometre. The boundary of one of these vast stations can be seen north of the river and is probably showing clearly because there are different grazing patterns on the adjoining farms. The boundaries themselves are generally marked only by barbed-wire fences.

Placenames are very revealing of an area's history and people and those in this part of the world are no exception. Some are anglicized aboriginal, such as Kynuna, Narangie and Cambeelo. Most, however, speak eloquently of the outbacker's nostalgia for their English or Scottish homelands. McKinlay, Woodstock, Middleton, Denbigh Downs and Windsor are all stations within the area of the image. Although ninety per cent of Queensland's population is Australian born, the early immigrants were almost exclusively of British stock, people who left their crowded little native island in search of a better future.

Placenames tell a similar story in the second image of Queensland, Figure 35, located south and west of Figure 34. The lake at bottom right is Lake Machattie, probably an aboriginal name; the station of Glengyle lies on its west shore, while Sandringham and Bedourie are smaller outstations that also appear on the area of the image (none of these stations is discernible). Cultural similarities apart, this image shows some striking contrasts with the previous one. Although only about 200 km (125 miles) further west, rainfall in this area is much less, and it is located on the eastern fringes of the great

Simpson Desert. Long yellow streaks of active sand-dunes slash across the image from the northwest, vivid evidence of the encroaching desert.

Even though it is almost in the centre of the continent, 1000 km (625 miles) from the sea, Lake Machattie is only 73 m (240 ft) above sea-level, and forms part of the great inland basin of Lake Eyre. It obtains its water from the creeks draining the slightly wetter Selwyn Hills to the north (Fig. 34 shows part of this area), and it is this water that sustains the vegetation showing up in bright red along the creeks. Floods during the southern summer rush down from the hills and fill the creeks with water for weeks at a time. When the waters come, mulga scrub – grasses, acacias and shrubs – flourishes, and provides grazing for the stations in the area. The image was obtained in July, during the southern winter. At that time, the lake still had some water in it, but towards the end of the year, before the rains come, it is usually a dry expanse of salt.

Intuitively, one might think that the dunes that dominate this image are aligned parallel to the prevailing winds. Until recently, this was in fact the accepted theory, but it has now been suggested that their origin is more complicated, because the prevailing wind direction changes with the season. Rather than being simply parallel to one seasonal wind, it is thought that the dunes are aligned in such a way as to average out seasonal differences. Whatever the details of their origin, the dunes seen here form only a part of the vast fields of dunes that cover the Simpson Desert. Charles Sturt in 1845 was the first European to reach the centre of Australia and discover the Simpson Desert, but it says much about the remoteness of the region that the first crossing and exploration of the Simpson was not completed until 1939.

Lake Machattie, then, lies on the edge of a great desert, and agriculture here is carried out at its limits. True deserts, by definition, are places where nothing can grow. Although most of the great deserts of the world support tiny groups of nomadic tribesmen, living frugally off whatever they can find, for millennia deserts have been synonymous with all that is hopeless and sterile. Throughout history, however, men have dreamed of making the deserts bloom, and in one or two places they have succeeded. Figure 36 shows one remarkable attempt, a startling cluster of centre-pivot irrigated fields in the heart of one of the most desolate regions of the world, the sand sea of Sarir Calanscio, Libya. How the irrigation scheme came into existence in the midst of this wilderness is a long and complex story,

involving characters as diverse and improbable as the Texas oil billionaire Nelson Bunker Hunt, and the eccentric and unworldly leader of Libya, Colonel Muammar al-Qaddafi. Working in consort with British Petroleum, Hunt's oil company made a major oil discovery at Sarir in 1961 which eventually became one of the most important oil-fields in the whole of Africa, exceeding even the massive fields of Texas. A pipeline was constructed from Sarir to the coast and revenues from oil exports started to flow into the country.

One of the more enlightened projects that was financed from the export of oil was a scheme to try and extend the area of agricultural land in this predominantly desert country, and help make Libya more nearly self-sufficient in food. No amount of money, of course, will make crops grow where there is no water, but fortunately the Libyans were able to tap a major aquifer, or stratum of water-bearing rock, which was found to underlay much of eastern Libya, notably around the Kufrah oasis (to the south of this image). Water was pumped to the surface and used in centre-pivot irrigators similar to those in Figure 33, first at Kufrah but later at Sarir. Initially, the irrigated circles were used to grow alfalfa to feed Barbary sheep. This scheme proved extremely expensive, however, and eventually grain crops were tried. As Figure 36 shows, only a small proportion of the circles were carrying a crop when the image was acquired in June 1983. Information is difficult to obtain from Libya, and it is not certain what the present status of the scheme is.

Watering the desert to make it fertile might seem an excellent objective, but there are some practical problems which are typical of all such schemes in desert areas. First, the aquifer that is being tapped consists of 'fossil' water that is not being recharged. The water that is being used for irrigation fell as rain millennia ago, and will not be replaced. One practical result can be that the water-level in the aquifer will drop below the level where it can be reached by the traditional rope-and-bucket wells used by nomadic tribesmen who also depend on this water. More seriously, in any desert area where crops are irrigated artificially, there is no flow of water to carry away salts that are leached out of the soil. Since the water is constantly being evaporated at the surface under the hot desert sun, unless the scheme is very carefully managed salt deposits can quickly build up in the top few centimetres of the soil, and eventually kill off growing plants. This has been the unfortunate outcome in many desert areas where optimistic irrigation schemes have ended in bitter disillusionment.

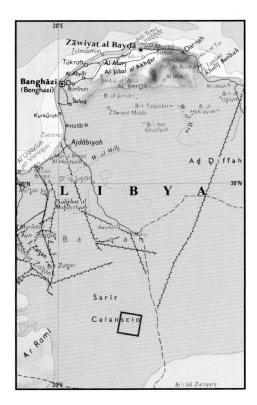

36 Double rows of centre-pivot irrigation discs reveal the hand of man in the heart of the Sahara: Sarir Calanscio, Libya (Shuttle 7, natural colour). Image scale 1 cm = 3 km; map scale 1 cm = 80 km.

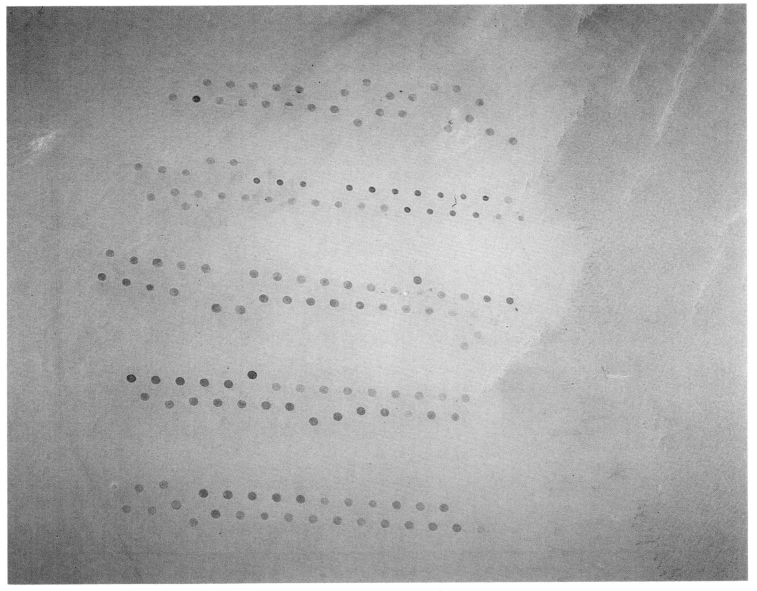

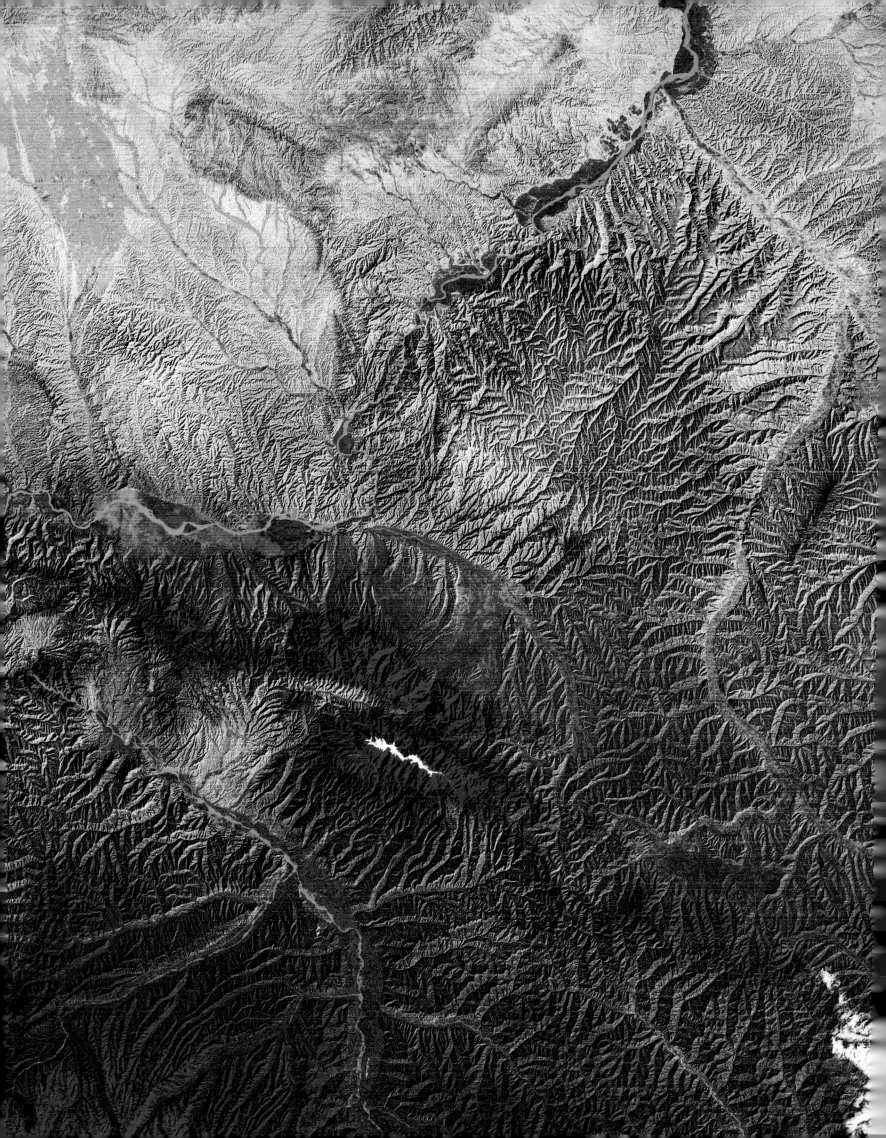

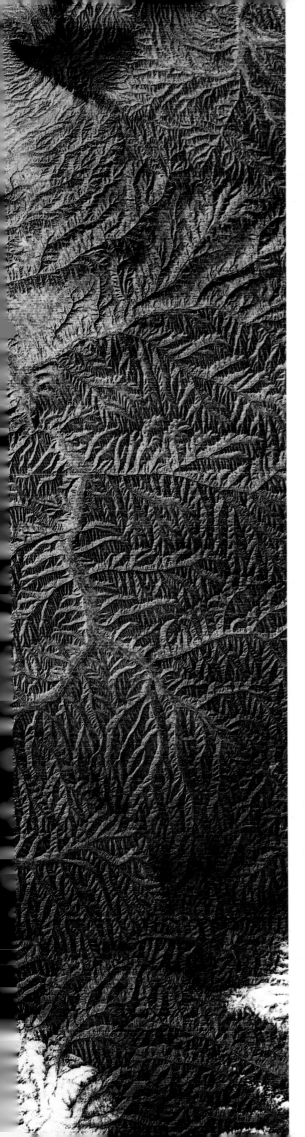

37 Intricate patterns of dry gullies nibble away at soft rocks in the upper reaches of the Yellow river, China (Landsat, false colour). Image scale 1 cm = 5.5 km; map scale 1 cm = 60 km.

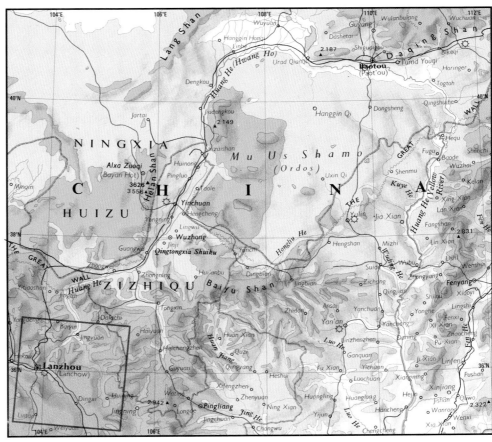

Libya does not have a single river within its boundaries, and the lack of water is the biggest single obstacle to its development. China, by contrast, suffers from an excess of water. The Huang He (Yellow river) is also known as 'China's sorrow', since it has been responsible for untold misery and loss of life throughout China's long history. In the flood of 1933, for example, more than 3000 villages were submerged and 3,600,000 people were drowned or injured. The Yellow is the second-longest of China's rivers (4845 km, 3030 miles), and drains from the Tibetan plateau through the north China plain to the Yellow Sea. The image selected to illustrate it (Fig. 37) shows a relatively tranquil area of the upper basin, but it is here that the trouble starts.

Two things are clear from the image: the headwaters of the river are rather dry and barren, and the rocks are easily eroded. The small patches of red along the river itself are evidence of agriculture, predominantly rice growing, and the red on the hilltops indicates that these too have a cover of dense vegetation. The rest of the area is scrub-covered or bare. To the geologist, the intricate filigree pattern of dendritic gullies covering the image speaks of rapid erosion of rather weak materials, in this case the fine-grained, wind-deposited silt known as loess. Being soft and fine-grained, the silt is easily eroded and carried off by the river, to which it gives a distinctive muddy yellow colour, the origin of the river's name.

Because the area is so dry and sparsely vegetated, any rain that falls on the hills runs off immediately, and the river rises and floods quickly. It is not the lack of vegetation and rapid run-off that make the Yellow river floods so lethal, however, but the loess. Vast amounts of it are borne in suspension by the river, floodwaters carrying up to 500 kg (1100 lb) of silt per cubic m (35 cubic ft). Not surprisingly, the Yellow ranks as the world's muddiest

river. As the gradient of the river decreases when it enters the lowlands near its mouth, the sediment is laid down again, raising the level of the river-bed. Left to itself, the river would shift its course frequently, flowing in shallow channels and gradually depositing an even layer of sediment over a wide area. For many centuries, however, the Chinese have tried to confine the river to its course by building high banks, thus guaranteeing catastrophe. Over the millennia, the river has deposited so much sediment along its constricted channel, and the banks have been raised to such an extent, that the water-level sometimes stands as high as 10 m (33 ft) *above* the surrounding countryside. When the river rises, there is always a great danger that it will overtop its banks and create a breach, pouring out over the flat low-lying surrounding region. This is exactly what happened in 1933. More than 110,000,000 people live in the basin of the Yellow river, so its seasonal behaviour is critically important to them.

Of all the natural hazards to which the Earth is subject, forest-fires are perhaps the most terrifying. In the right conditions a fire can become a spreading inferno that can destroy thousands of square kilometres of woodland in a matter of hours, sometimes engulfing populated areas as well. Every year, such fires sweep through the dry, wooded hills of California, Australia and many other places, leaving only charred wasteland behind them. Figure 38 illustrates just how extensive such forest-fires can be. Here, three major fires, fanned by a fresh southerly wind, were imaged in July 1975 as they blazed through remote pine forests near the Mackenzie river in Canada's North West

38 Plumes of smoke curl up from forest-fires blazing near the Mackenzie river, North West Territories, Canada (Landsat, false colour). Image scale 1 cm = 5.5 km; map scale 1 cm = 150 km.

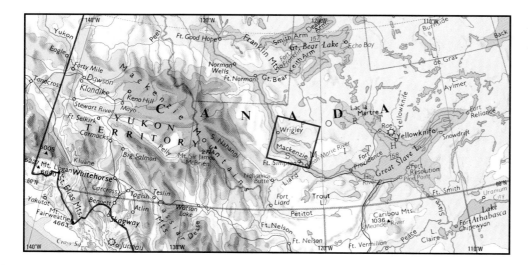

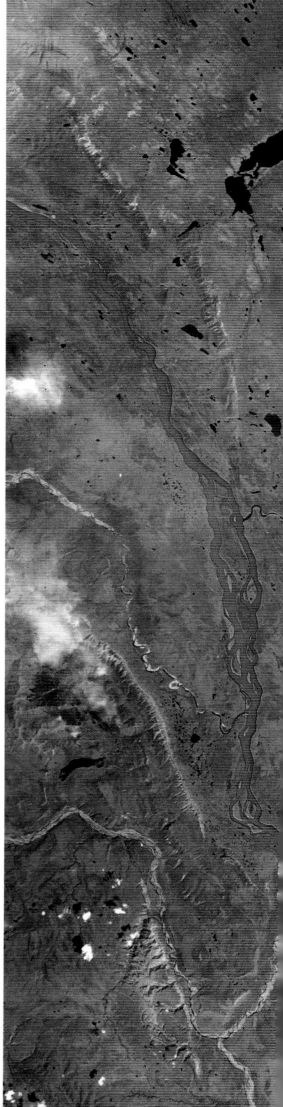

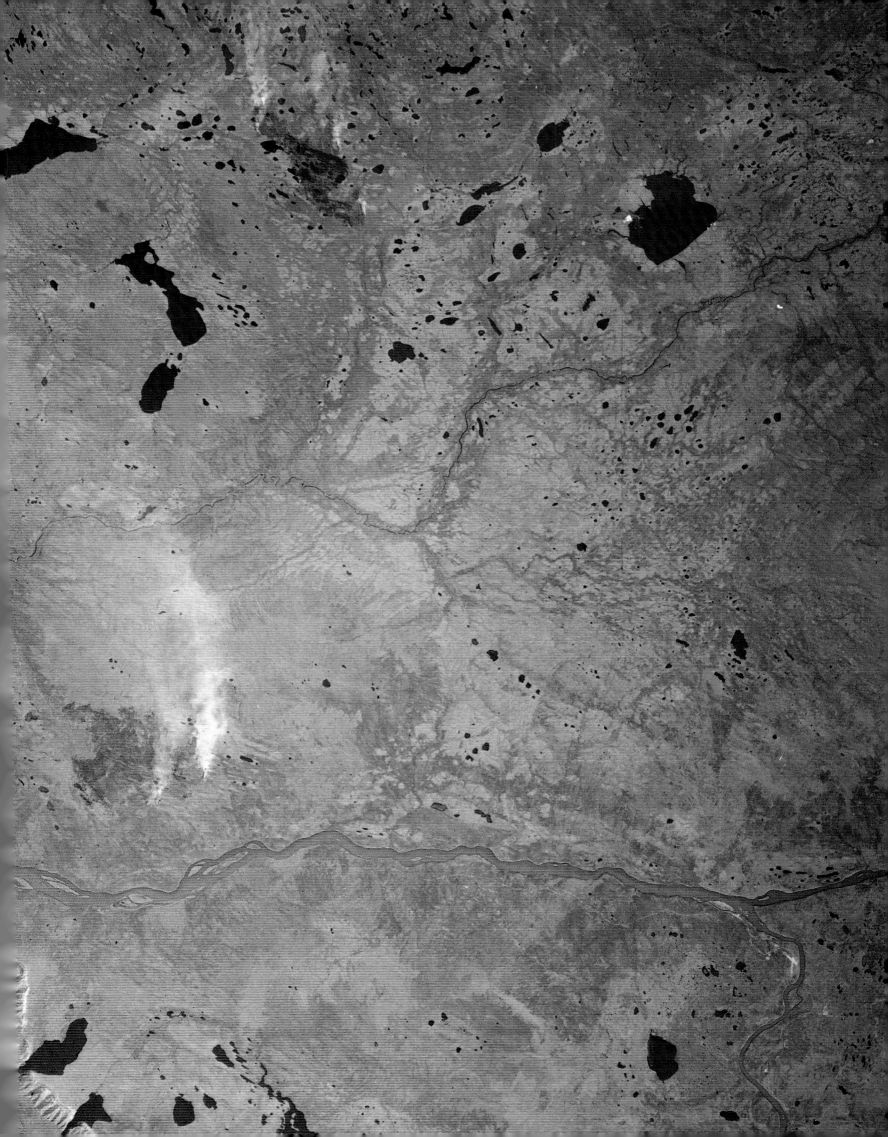

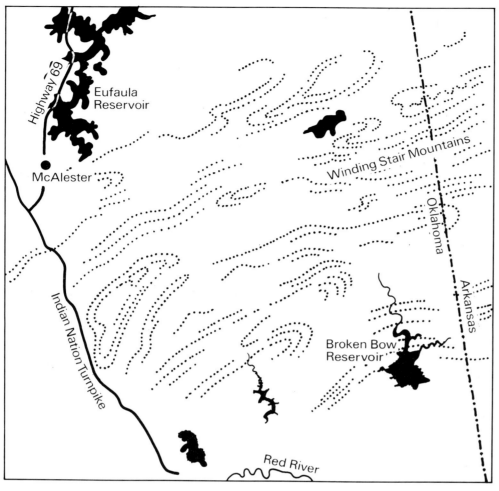

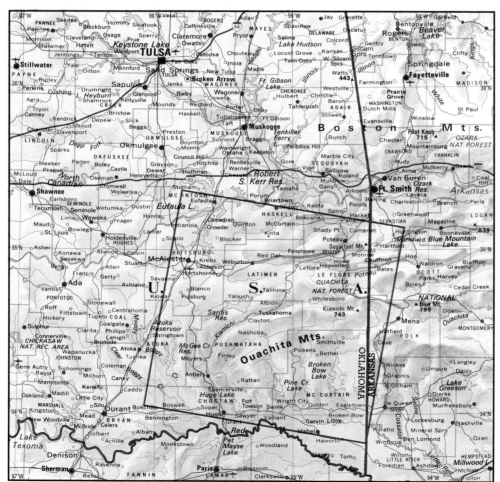

Territories. The fires had already reduced several hundred square kilometres of woodland to ashes, and were still well alight when the satellite came overhead, as the distinct plumes of smoke reveal.

The natural human response to a fire is to try to put it out. And it is equally natural that stringent precautions should be taken to prevent fires from being started. Heavy propaganda is aimed at careless cigarette smokers, hikers building camp-fires, and so on. In some parts of the world, forests are actually sealed off from visitors in dry weather, and draconian penalties imposed on trespassers.

While such precautions are understandable, and easy to sympathize with, they can sometimes be overdone. After decades of the most scrupulous fire-prevention policies, the United States forest service has realized that fires actually play an important part in the ecology of forests. If fires are rigorously prevented, the whole character of the forest departs from its 'natural' condition, and it eventually becomes senile. When a fire sweeps through a mature natural forest, the larger trees tend to survive, with scorched branches and foliage, but saplings and undergrowth are destroyed, thinning out the trees and giving better conditions for the survivors. The ashes from the fire also provide valuable fertilizer for the soil. Thus, while catastrophic accidental fires are still strenuously avoided in American National Forests, the policy now is to start occasional *controlled* fires, whose timing and spread can be managed, and thus allow the forest to regenerate itself naturally.

Satellite images provide an excellent means of monitoring the progress of large forest-fires, some of which may burn for weeks. Apart from simply identifying and quantifying the areas affected, they can assist in pin-pointing potential trouble spots. The fire blazing near the southern bend in the Mackenzie river, for example, may be burning itself out, since the 'hot spots' from which the smoke plumes are rising are on the southern, upwind side of the affected area. Thus, the wind is carrying and spreading sparks over already burned ground. In the northernmost fire, on the other hand, the hot spots are on the downwind side, so sparks can be blown into unaffected areas, spreading the fire.

The North West Territories are exceedingly sparsely populated, so fires like those in the image present little threat to human

39 PREVIOUS PAGE Clear-cut felling formed this patchwork design on the forested hills of eastern Oklahoma, USA (Landsat, Thematic Mapper, false colour). Image scale 1 cm = 4.5 km; map scale 1 cm = 25 km.

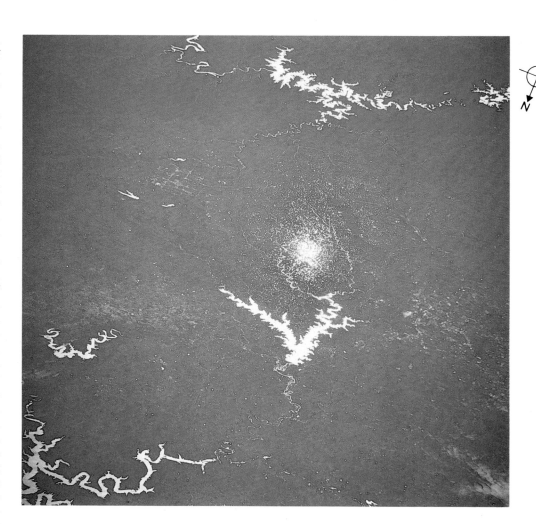

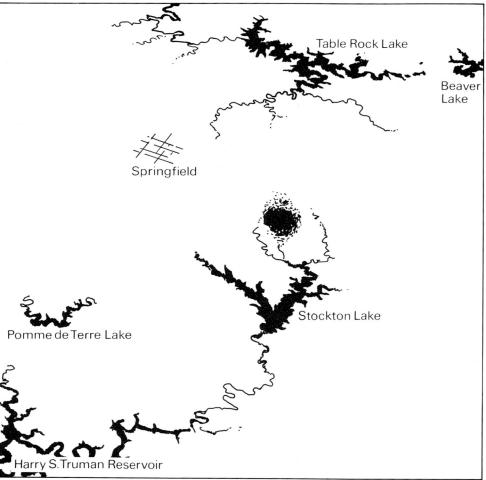

Table Rock Lake

Beaver Lake

Springfield

Pomme de Terre Lake

Stockton Lake

Harry S. Truman Reservoir

40 A ball of light marks the sun's reflection in numerous pools and rivers near Springfield, Missouri, USA (Shuttle 9, oblique view looking south, natural colour). Image scale 1 cm = 8 km; map scale 1 cm = 25 km.

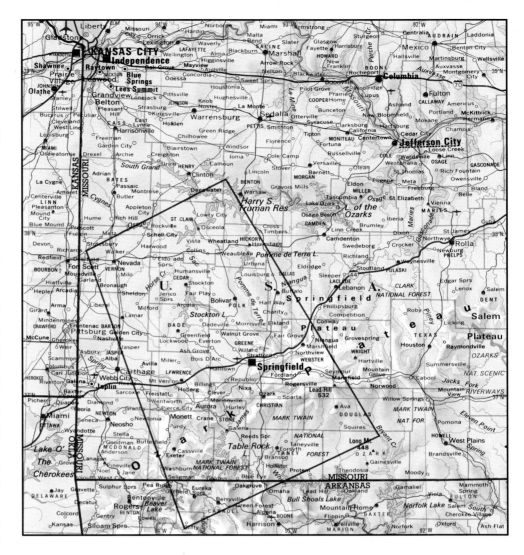

life, and can usually be left to burn themselves out. The overall orange tone of the forest in the image is distinctly mottled, with patches of lighter and darker colour. Many of the lighter patches probably represent old burned areas, where new growth is well-established. Eventually, the blackened areas left by the active fires will also be covered by new, healthy vegetation, and the animal community that was destroyed or fled will re-establish itself, completing the natural cycle.

The rate at which trees are being felled for agricultural land, timber and paper-pulp is a more serious threat to the world's forests than forest-fires. The most striking feature in the image of Oklahoma (Fig. 39) is the chequered effect created by clear-cut felling on a massive scale – the red forested areas stand out strongly, contrasting sharply with the blue of the square plots where trees have been felled. The scale of the operation might suggest that lumber projects were of prime importance to the economy of Oklahoma, but the glaring environmental impact revealed from space does not reflect such economic significance. Petroleum products are by far the biggest money earner for the state, while wheat is the leading cash crop.

At least 16,100 sq km (6200 sq miles) are commercially exploited for timber in Oklahoma. Pulp and paper industries were introduced in the 1970s adding further demand for wood to the established enterprises. One of the reasons why the timber industry is attracted to this state is that both hardwoods and softwoods are available in close proximity to each other. Lumbering as seen here is a highly attractive commercial proposition. Compared with the cost of a mining operation, or the investment required in arable farming, clear-cut felling needs minimal financing as there is often no attempt to replant cleared areas – the trees are simply felled, hauled to a mill, sawn up and sold. In the long term this practice cannot continue as woodland resources will become exhausted and there are already vigorous environmental groups opposing clear-cut felling.

Fortunately, some 1400 sq km (540 sq miles) of Oklahoma woodlands come within the preserve of the Ouachita National Forest. One section of this appears in the bottom right corner of the image, its eastern limit precisely defining the straight line which forms the state

boundary with Arkansas. Another section of the National Forest extends from Arkansas along the Winding Stair Mountains at centre right of the image (appearing deeply ridged). Here, as in the state park surrounding the Broken Bow Reservoir (lower right), tree felling is officially controlled, and the natural forestation shows in solid red with only a few lighter areas, contrasting sharply with the patchwork pattern of exploited forest.

Just snaking into the bottom of the image is the Red river, forming the southern boundary between Oklahoma and Texas. The Indian Nation Turnpike is clearly visible running approximately north-south down the western edge, and the symmetrical smooth curves of the freeway intersections on highway 69 passing south over the Eufaula Reservoir to McAlester can be traced from space as if on a planner's drawing-board. The image reminds us, however, that vast amounts of land are consumed in the interest of conveying us from one fast lane to the next with the minimum of deceleration and maximum convenience.

Looking south over Missouri towards Table Rock and Beaver lakes on the state

line with Arkansas, Figure 40 not only provides an example of the visual feast awaiting space travellers, but also shows how man can transform his environment, for better or worse. The low December sun sparkles brilliantly off countless lakes and pools, while other details are subdued in the glare. Several large, gleaming lakes and myriads of small pinpoints of light emphasize how much standing water there is in this area of the USA. The glistening effect seen here partly reflects the fact that this image was acquired in winter, when all drainages were full from recent rains, but in addition the area has always been a rather wet one. Much of the country south of the Missouri river was swampland until comparatively recently.

The lakes visible in the image are all artificial and have contributed substantially to the sweeping changes in the environment that have taken place in Missouri. At bottom left, the Harry S. Truman Reservoir snakes its way for at least 32 km (20 miles) along a drowned river valley, while the dam responsible for the end of Stockton Lake (centre) can be discerned as it crosses another valley. These reservoirs form part of an elaborate drainage scheme that has helped to turn southern Missouri into a rich farming area, benefiting from one of the most extensive areas of fertile alluvial soils in the USA. In the nineteenth century, Missouri was a major cotton-growing state, but the fields were confined to the better-drained parts, away from the swampy Missouri river floodplain. Drainage of the swamplands has not only brought more land into production, but has also given more scope for diversification. The primary crop is now soya beans, with cotton and rice in second and third places.

Although construction of the reservoirs and draining of wetlands has brought great benefits to Missouri, the environmental cost is significant. Swamplands all over the world are being drained to provide more agricultural land, and where this happens many existing plant and animal species are permanently displaced, and their habitats lost for ever. Although the new lakes themselves provide attractive recreational facilities, and habitats suitable for some kinds of wildlife, the loss is a serious one. The world is not short of lakes, but it is getting short of wetlands. In the Missouri image, the vast extent of the reservoirs gives an impression of the scale of the problem. Fortunately, the situation in Missouri is not as serious as in some other areas. In Florida, for example, the drainage of wetlands has been so extensive in the last decade that it is beginning to rate as an ecological disaster area.

The loss of specific kinds of habitat, such as Florida's wetlands, is merely one result of the most universal and intractable pressure on the Earth: the demands imposed by a rapidly-increasing world population. Population pressure, in fact, is probably the root cause of most of the environmental problems that the Earth faces today and nowhere are these problems more obvious than in Japan. Figure 41 is dominated by the serene elegance of Mt Fuji, at 3776 m (12,390 ft) the highest mountain in Japan, and deservedly renowned for its sweeping symmetrical profile. The Japanese venerate the mountain: there is a shrine on top, visited by thousands of pilgrims each year, and the mountain is protected from despoliation by its location in the Fuji-Hakone-Izu National Park. Around its base are a number of lakes, particularly valued by the Japanese with their highly developed, sometimes rather rarified aesthetic sensibilities. One of them, Kawaguchi-ko, is especially treasured for the inverted reflection of Fuji that can be seen on the tranquil surface of its water. Although so highly prized for its peacefulness, Fuji is technically an active volcano and last erupted in 1707.

North and west of Fuji extends a jumble of deeply-gullied forested hills. Too steep for agricultural purposes, such hills occupy a high proportion of Japan's land area. As a result, farms, villages and towns are squeezed onto the narrow tracts of flat land along valley floors and fringing the coast. Several major coastal towns can be seen both east and west of Fuji, and the valley to the north is also heavily built-up. The pressure on land is extreme. It reaches its peak in the area at the top of the image, where the grey sprawl of the Tokyo-Yokohama-Kawasaki metropolitan area stretches away into the distance. Over 22 million people live here; in parts of Tokyo the population density exceeds 15,000 per sq km (39,000 per sq mile) and in Yokohama and Kawasaki it is typically over 5000 per sq km (13,000 per sq mile). Such large numbers of people require enormous quantities of food, water and energy, and provision of these is a major strain on the economy.

Most of Japan's energy has to be supplied from overseas, and the amount of useful agricultural land is so restricted that what there is has to be intensively utilized in order to decrease the nation's dependence on imported foods as much as possible. Japan has been outstandingly successful in building up export industries to pay for imports, but this success has had its own price. Apart from the sheer ugliness of the Tokyo area, much of Japan suffers from a very high level of industrial pollution. Although more stringent controls have been introduced in recent years, the country has provided some of the world's classic examples of industrial pollution, such as the poisoning of a whole community who ate fish from a bay contaminated by effluent from a local

factory that included high levels of mercury. In the image, the dockland district of Yokohama is clearly discernible, but most of the Tokyo area is obscured by a hazy pall of smoke and fumes.

Apart from such artificial pressures, the Tokyo area is also subjected to severe natural stresses and strains. Frequent earthquakes rock the city, and rigorous building codes have been drawn up to minimize damage and casualties in the event of powerful tremors. And, situated so close to a major volcano, the urban areas would all be severely afflicted if a large eruption were to take place. One might suppose that the combination of natural and artificial pressures would limit Japan's progress, but the reverse is true: the country goes from strength to strength each year.

Figure 42 shows another part of the world where the topography is dominated by active and recently active volcanoes, but one which is almost entirely uninhabited. Located on the frontier between Chile and Argentina, most of the area is between 3000 and 4000 m high (10,000 to 13,000 ft) and some of the peaks reach over 6000 m (20,000 ft). It is one of the highest and driest deserts in the world, where human life is confined to one or two small mines and a handful of hamlets where springs provide enough water for llamas to graze.

Because it is within the tropics, and also at high altitude, the climate is extreme. The intense sunshine can raise daytime tem-

41 A serene volcano and sprawling cities: Mt Fuji and the Tokyo-Yokohama-Kawasaki metropolis (Shuttle 2, oblique view, natural colour). Image scale 1 cm = 10 km; map scale 1 cm = 25 km.

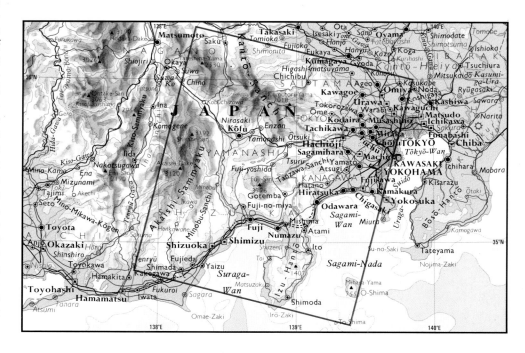

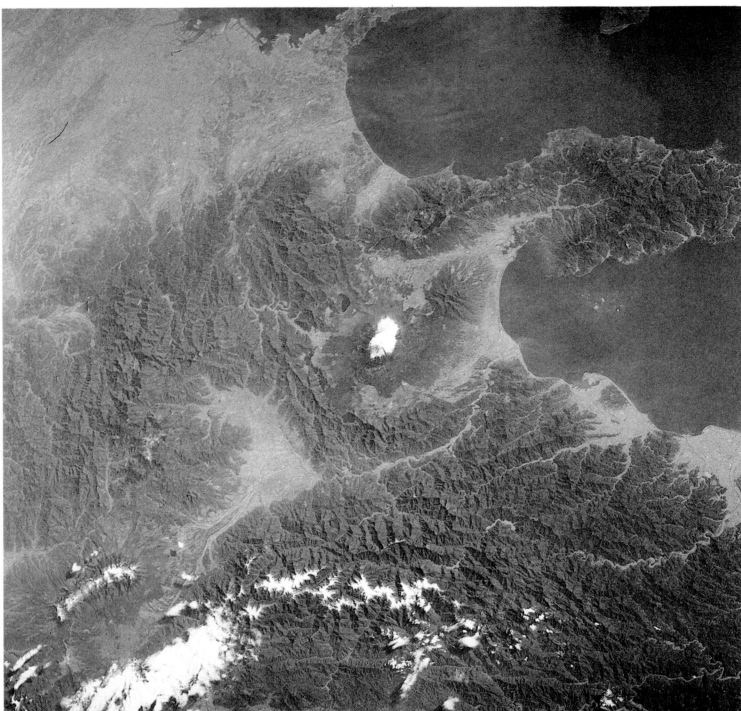

peratures to 30°C (86°F), but the same night it may plummet to −15°C (5°F), as the day's heat is radiated away into space through the thin, dry air. Coupled with the temperature range is extreme dryness – the region contains some of the largest salt lakes in the world, where any small amounts of rain or snow are rapidly evaporated. Perhaps the worst element of the climate, however, is the savage wind. The ragged plumes visible in the image are not the columns of volcanic eruptions, but dust plumes whipped up from the surface of the desert by vicious winds. Such winds blow unrelentingly for most of the year, always from the northwest, usually only

strongly, but sometimes, as in the image, at gale force.

In the long term the consequences of the consistent wind patterns are quite profound. The area is so high and dry that wind rather than water is the principal agent of erosion. Deep gullies and grooves have been excavated in soft volcanic rocks (visible right centre), and the dust is whirled away downwind. Locally, a thickness of more than 100 m (330 ft) of ash has been blown away, producing unusual valleys and deflationary hollows.

Such wind erosion is highly selective, partly because it affects some materials much more than others and partly because

the topography has to be right. Clearly, soft silts and ashes will erode much more quickly than hard lavas. Careful inspection of the image shows that dust plumes of two distinct colours are being produced, red and white. The red plumes (top) are being whipped up from rust-coloured silty sediments, probably deposited originally as fine muds in lake basins, while the white plumes are derived from the surface of salt lakes, and consist of tiny particles of evaporite minerals such as gypsum. Although there are many salt lakes on the image, plumes are rising from only a few places. This is because they will only occur where the salt surface is dry enough for the

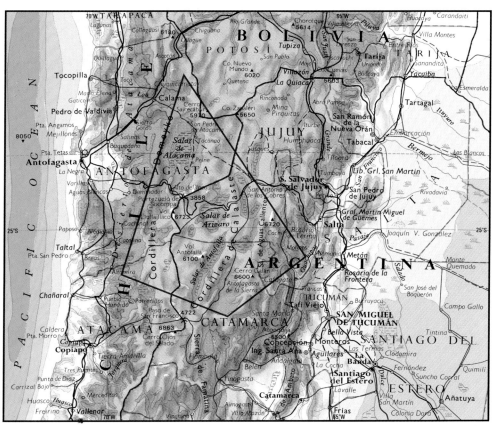

crystals to be picked up, and where the local topography is such as to focus the wind to raise the dust. Apart from blowing horizontally, the wind must have an upwards component too. These conditions are typically found downwind of mountain ridges, where the churning, revolving eddies of air can have quite strong upwards motions. Several of the plumes in the image are in fact located in exactly this kind of position relative to ridges.

Although the area in the image is too desolate for dust storms to cause serious problems, elsewhere, especially in agricultural areas, they can be devastating, since they are capable of whipping away the fertile topsoil, leaving nothing for crops to grow in. This usually happens when an area has been too intensively farmed, so that the soil is easily broken up and transported. But in some circumstances dust storms can serve a useful purpose. The dust that is entrained may be carried for hundreds or thousands of kilometres downwind. In the case of the storm in the image, the dust will eventually fall to earth over the Argentine pampas as a barely measurable film of powder, possibly visible on polished surfaces like car windscreens. Over the years, however, this input of airborne sediment represents a significant contribution to the soil of the pampas, enriching and fertilizing it.

42 Vicious winds generate swirling dust-storms along the Chile/Argentina frontier (Shuttle 8, oblique view, natural colour). Image scale 1 cm = 10 km; map scale 1 cm = 80 km.

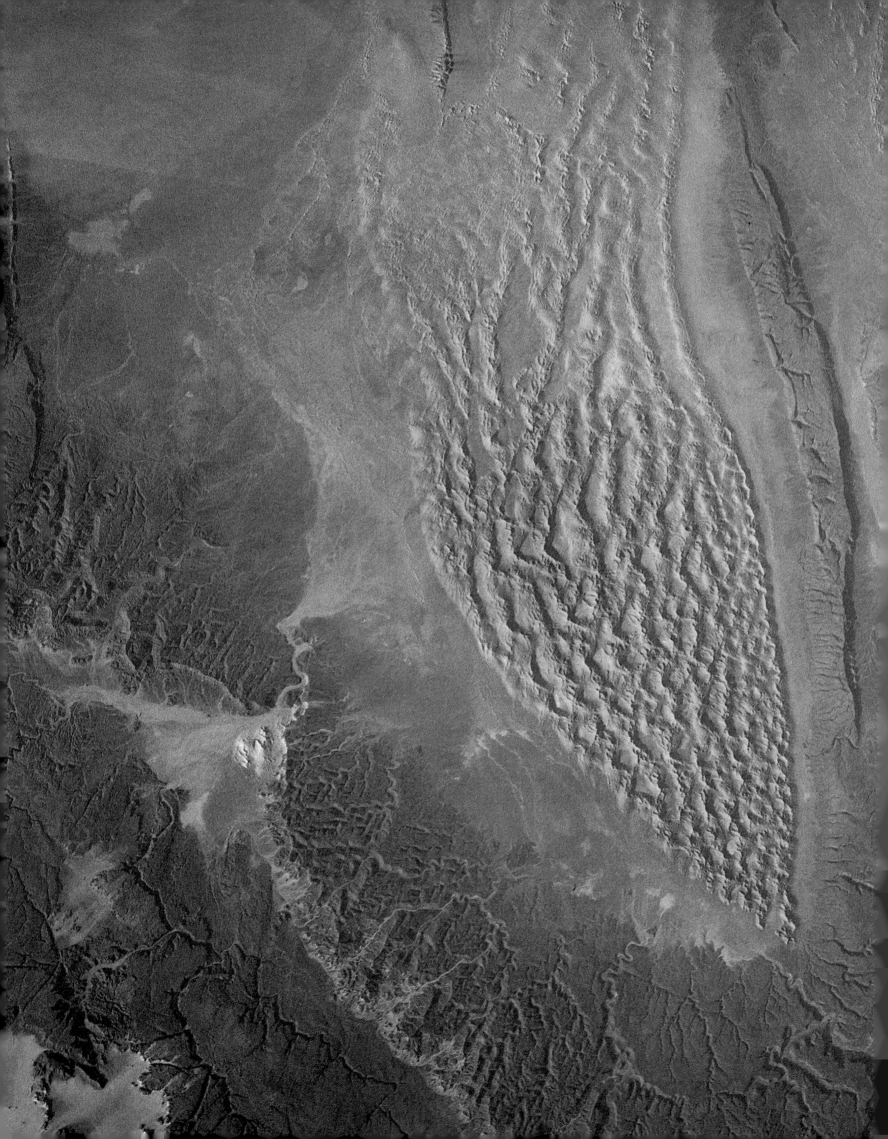

5 The Restless Earth

Of the nine planets that orbit the Sun, life has evolved on only one, the Earth. There are many reasons why only the Earth should be so favoured, but one of the key factors is that the Earth itself is alive, geologically speaking. With the exception of Venus, the other planets are merely frigid balls of ice and gas, or inert lumps of rock, like the Moon, which have remained essentially unchanged for hundreds of millions of years. Spacecraft missions are beginning to hint that Venus, Earth's nearest neighbour in space, may not be as inert as the other rocky planets, but its searing temperatures ensure that its surface will be a sterile, arid wasteland. On the Moon, the footprints left by the Apollo astronauts will still be clear and sharp when all of us are long dead. By contrast,

43 A dagger-like formation of star sand-dunes pierces the stony wastes of the Sahara, Grand Erg Oriental, Algeria (Shuttle 2, natural colour). Image scale 1 cm = 4 km; map scale 1 cm = 80 km.

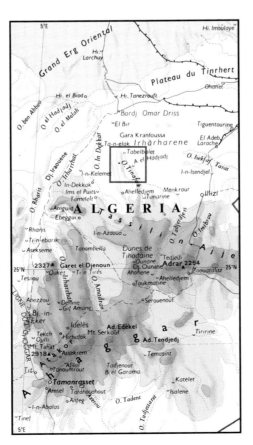

so intensive are the surface processes on the Earth that footprints will usually be erased by wind or water within a few hours or days. The mountains and hills that seem so unchanging to us are, geologically speaking, very short-lived – under tropical conditions, a 3000-m (10,000-ft) volcano can be erupted and entirely eroded away within five million years.

So rapid are the processes that operate to sculpt the face of the Earth that many of them are perceptible even on the scale of a human lifetime, and even those that are imperceptibly slow often reveal their work in ways which are beautifully encompassed in space images. Figure 42 showed the wind at work as an agent of erosion in the Atacama desert, one of the great deserts of the world. Figure 43 shows the results of sediment deposition by the wind in another great desert, the Sahara. Thrusting into the heart of the Sahara is the Tifernine field of sand-dunes, the southern part of the vast sand sea that constitutes the Grand Erg Oriental of eastern Algeria. Apart from small groups of nomadic tribesmen, few people have ever seen these dunes on the ground, since they lie in one of the most desolate parts of the Sahara. The nearest oasis is over 100 km (62 miles) distant, though some maps show a well at Tabelbalet (near the top centre of the image, but not distinguishable) and a caravan trail winds along the valley separating the two main dune fields. The trail can only rarely be used, however, since there is not the slightest hint of a track in the image, whereas well-used trails in desert areas are usually conspicuous (as Fig. 13 shows clearly). Such travellers as venture along this route would see little of the beauty that the image conveys. All they would see would be the lines of high dunes marking the edges of the dune fields, a daunting and discouraging prospect. No rivers flow in the area, and there is no vegetation. If camel caravans are to survive in such terrain, they are absolutely dependent on finding wells.

The texture of the dunes in the image is highlighted by the evening sun, revealing the structure of the field. Where dune fields are built up by winds blowing consistently from a single direction, crescent-shaped dunes known as barchans are usually formed. The steep convex side of these dunes always faces into the prevailing wind, while the tapering arms face downwind. In areas where there is no single prevailing wind direction, dunes tend to form more symmetrical shapes, known as star dunes. The Tifernine dunes show elements of both forms intermingled. Where the dunes extend furthest into the desert, they have many of the characteristics of star dunes, but barchan shapes are developed further north, with star-like features superimposed on them. The curves of the barchans suggest that the winds responsible for them blow from the west or southwest.

In desert regions, wind is often the most important agent of erosion, although the effects of flash floods following rare rain storms can be remarkably profound. Sharply incised gullies dominate the lower left corner of the Tifernine image and emphasize the contribution that water makes even to Saharan scenery. On the other side of the world, several orders of magnitude larger in scale, the winding ribbon of the Yangtze (Chang) (Fig. 44) demonstrates just how powerful water can be in shaping the face of the Earth.

The Yangtze is the longest river in Asia. It rises high on the Tibetan plateau and makes an enormous southward loop before turning east. It finally enters the sea near the great port of Shanghai, a total length of 5500 km (3400 miles). Of this, the

lowermost 1500 km (900 miles) is navigable by ocean-going ships, while more than 2000 km (1250 miles) is navigable by smaller vessels. Figure 44 shows part of the middle reaches of the river, in the province of Szechwan (Sichuan). Southwest of the area shown, the river flows through flat country, is nearly half a kilometre wide, and is easily navigable. Something of this aspect of the river's character can be seen in the bottom left-hand corner of the image, as it loops and winds along a broad valley. In the top right (near the city of Wanxian), however, the river has cut its way through resistant rock strata which have confined it into an extremely narrow channel. For tens of kilometres it flows through the terrifying Shangxia gorges, which cause serious problems for navigation. The gorges are rocky chasms with vertical walls up to 600 m (2000 ft) high; the fast-flowing river imprisoned within them is constricted to a width of less than 150 m (500 ft). So large is the volume of water surging into the gorges that the river becomes extraordinarily deep (over 152 m, 500 ft), making the Yangtze the deepest river in the world.

Impressive though the gorges of the Yangtze are, they are little known outside China, and pale into insignificance in comparison with the renowned Grand Canyon on the Colorado in the USA (Fig. 45). Everyone has heard of the Grand Canyon: it is almost synonymous with everything that is vast (and American) and has figured in countless photographs and postcards. Figure 45 covers the western part of the Grand Canyon National Park; the most frequently visited areas at Grand Canyon village and North Rim lie just off the right side of the image. Statistics

44 Longest river in Asia, the Yangtze here threads its way through fold mountain ranges to the Shangxia gorges, Szechwan province, China (Landsat, false colour). Image scale 1 cm = 5.5 km; map scale 1 cm = 60 km.

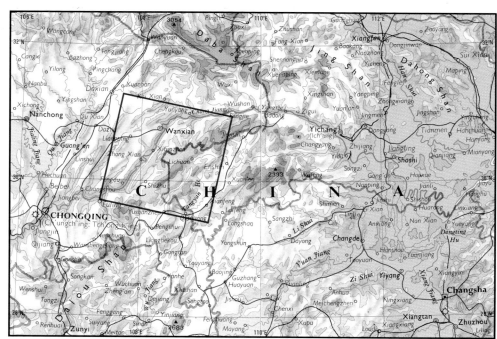

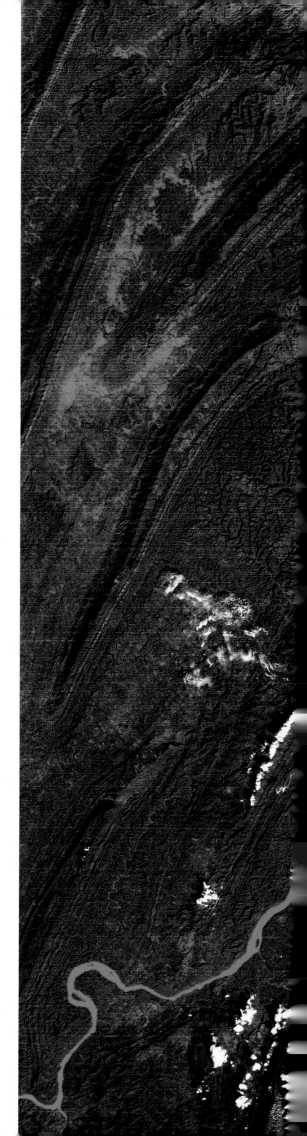

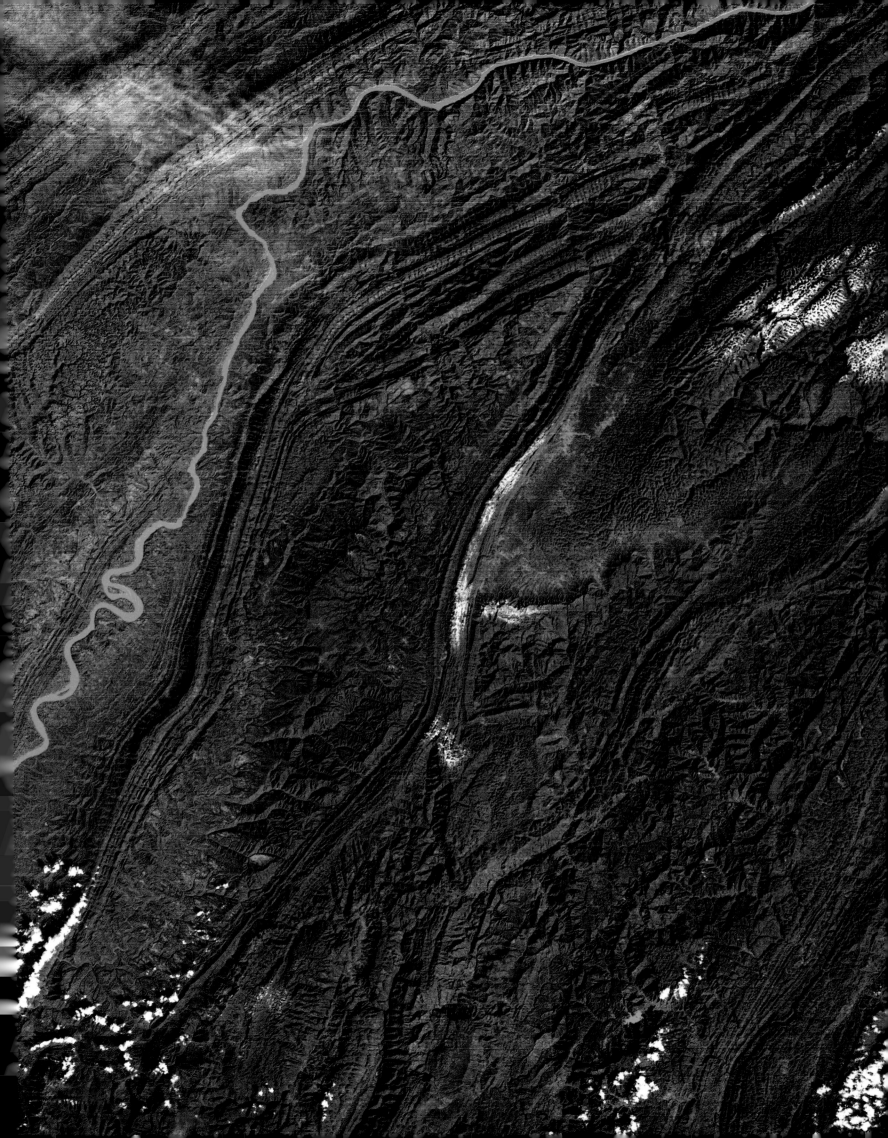

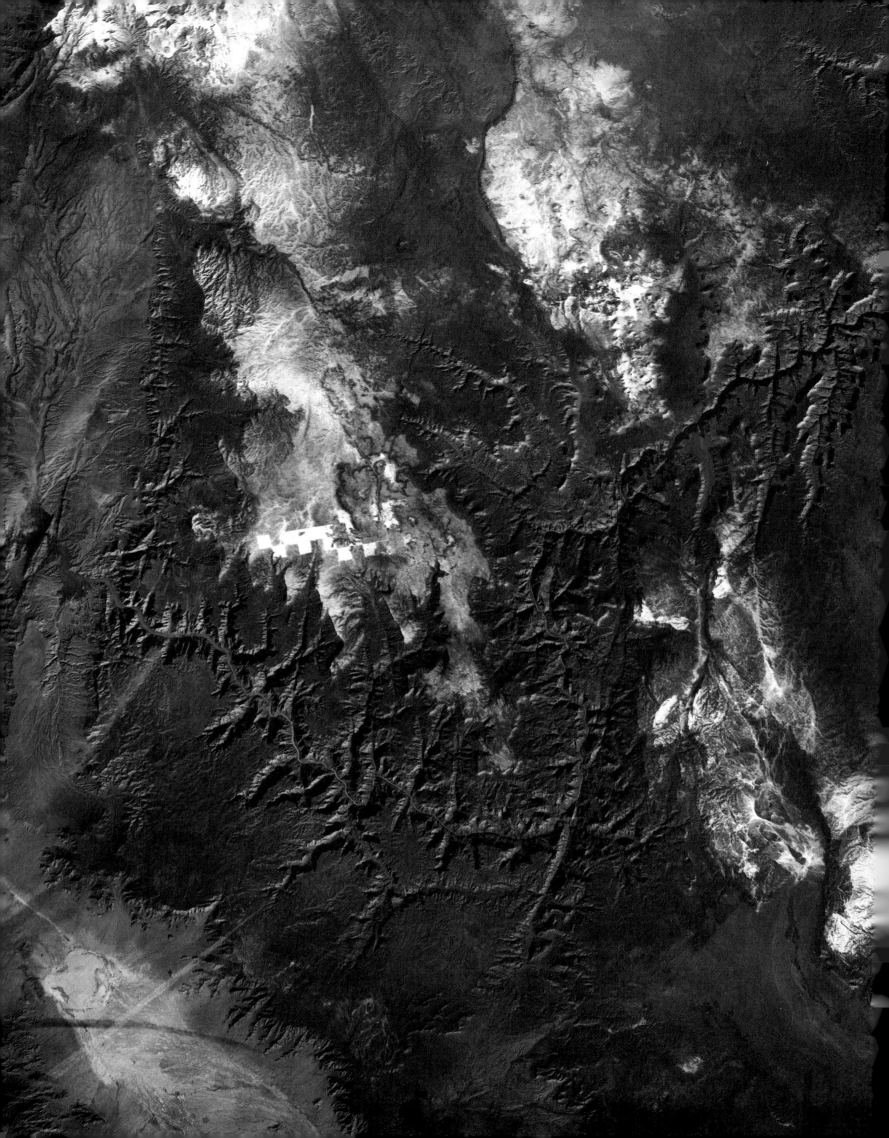

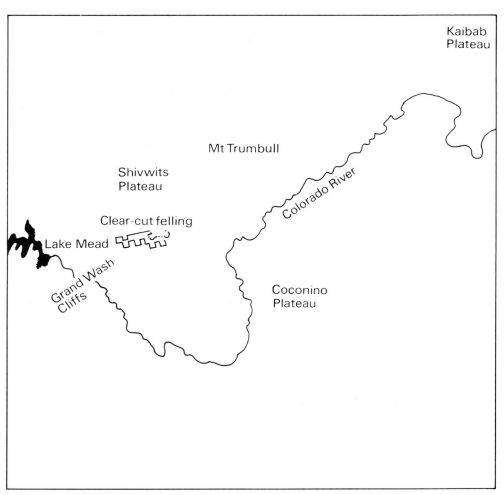

45 Snow mantles high ground along the western part of the mighty Grand Canyon, Arizona, USA (Landsat, Thematic Mapper, false colour). Image scale 1 cm = 5 km; map scale 1 cm = 25 km.

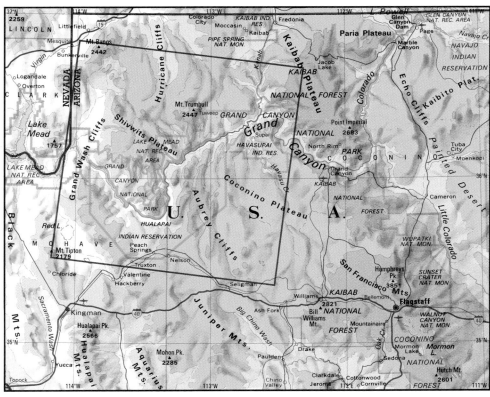

inevitably come into any discussion of the Grand Canyon, but they are so impressive that they are worth repeating. The canyon extends 349 km (218 miles) from Marble Gorge to Grand Wash Cliffs; it is only 6 km (less than 4 miles) wide at its narrowest point but 1620 m (5300 ft) deep, The national park along the canyon is 56 km (35 miles) across, but within it the Colorado river winds for 169 km (105 miles); and the North Rim is 2500 m (8200 ft) above sea-level, 400 m (1300 ft) higher than the South Rim.

The image selected to illustrate the canyon was acquired in winter and snow mantles the high ground of the Coconino, Kaibab and Shivwits plateaux. The low sun of a winter morning highlights the majestic cliffs facing south and east; north- and west-facing cliffs are deeply shadowed. On the right side of the image, where the gorge is narrowest, the Colorado river is lost in the shadows. (So steep is the canyon that some parts of its floor are perpetually shadowed, even in summer.) On the left side of the image, where the gorge is wider, the blue thread of the river can be seen winding its way beneath towering cliffs before entering Lake Mead, an artificial reservoir, on the extreme left. A row of neat white squares, each exactly one mile (1.6 km) across and clearly artificial, seem totally incongruous in comparison with the natural grandeur around them. The squares are areas where timber growing on the Shivwits Plateau has been clear-cut and where the snow is now lying evenly. In contrast, where timber is still standing, the snow cover rests unevenly on the foliage of the trees. Although such clear-cutting in an area of outstanding natural beauty might seem out of place, no laws are being broken, since the area lies just outside the boundaries of the national park.

A perceptive visitor to the Grand Canyon is likely to start wondering why it is so deep and steep. The first question is particularly interesting, because the Colorado river which cut the canyon is by no means an especially large river. In terms of its length and the volume of water flowing in it, it is a mere mountain stream compared with the Yangtze. Although there are some contributory factors, the fundamental reason why the canyon is so deep is that it was carved by the Colorado river *through land that was rising in its path*. The Colorado was flowing along roughly its present course when the Coconino and Kaibab plateaux were being elevated by geological forces; as the land slowly rose, the river kept on cutting down, trying to maintain its original gradient to the sea.

The steepness of the canyon is less easy

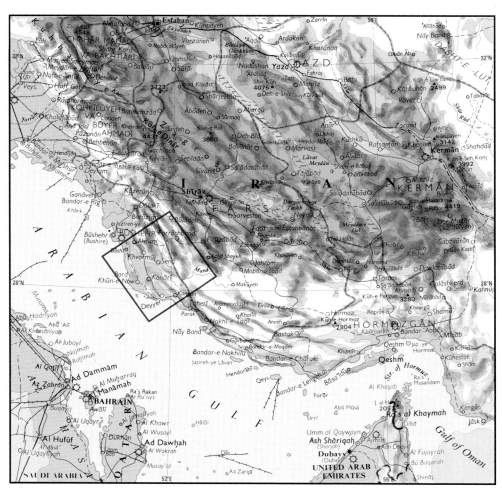

to account for. There are many canyons much deeper than the Grand Canyon in the Andes and Himalayas, but they are shrouded in obscurity not only because they are less accessible but also because their walls slope more gradually and therefore less impressively than those of their more famous counterpart. One reason for the canyon's dramatic form is that it is excavated through horizontal rock strata, notably the Coconino sandstone. This rock fractures perpendicularly to form continuous and spectacular cliffs along the whole length of the canyon, providing a perfect natural example of textbook geology. A less obvious reason is the length of time that erosion has been taking place. The Grand Canyon is fairly young, while the deeper canyons of Peru and north Chile, for example, are rather old (compare Figs. 19 and 50), initiated more than ten million years ago. With the passage of geological time, and slow but continuous erosion, these great canyons have developed gentle slopes.

While quite large vertical movements of the Earth's crust took place in the formation of the Grand Canyon, they did not create any structures that are immediately obvious in the image. In Iran's Zagros Mountains, by contrast, fold structures produced by compression of the Earth's crust are crisply and un-

ambiguously displayed (Fig. 46). Here, layers of sedimentary rocks laid down beneath the sea only some 60 million years ago have been buckled up into a series of folds, known as anticlines and synclines. Anticlines form the arches of the folds, while synclines form the corresponding basins between them. In the image, the anticlines form the conspicuous boat-shaped ridges, their crests stripped away by erosion. On the flanks of the anticlines, beds of softer rocks have been eroded away along gullies. More resistant rocks form the crests of small ridges, creating a jagged, saw-tooth effect (lower right). Anticlines are the stuff that oil company geologists dream of, as these formations may carry oil trapped beneath an impermeable layer. Iran contains many major oil-fields, some of them located in exactly this kind of structural setting. A minor example, the Kuh-i-Mand field, lies at the easternmost end of the fold nearest the sea in the image, though the workings are too small to be discernible.

Geologically speaking, the Zagros Mountains are extremely young since they affect newly-deposited sediments. They provide striking visual confirmation that the Earth is a dynamic, evolving planet and that *new mountain ranges will be produced in the future*. By contrast, the Macdonnell Ranges of Australia provide a

46 Young fold mountains of the Zagros range along the Gulf coast of Iran (Shuttle 2, natural colour). Image scale 1 cm = 8 km; map scale 1 cm = 70 km.

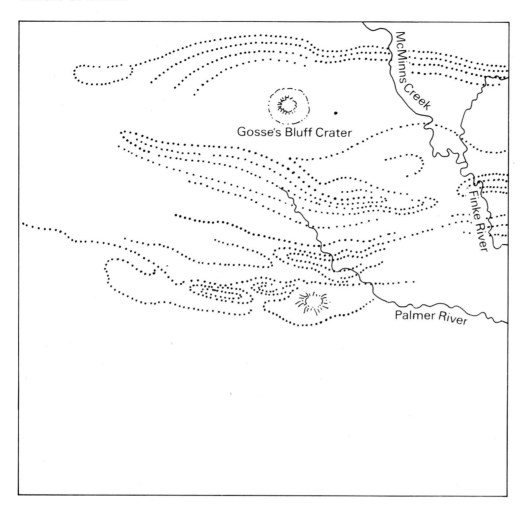

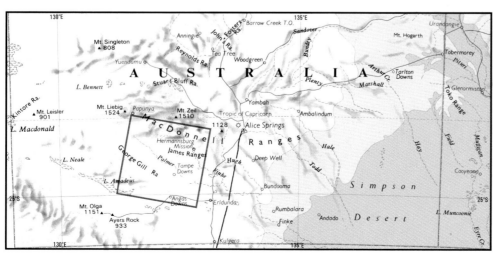

47 Ancient fold mountains form the Macdonnell Ranges, central Australia (Northern Territory). Gosse's Bluff Crater, where a small asteroid collided with the Earth, is at top centre (Landsat, false colour). Image scale 1 cm = 5.5 km; map scale 1 cm = 80 km.

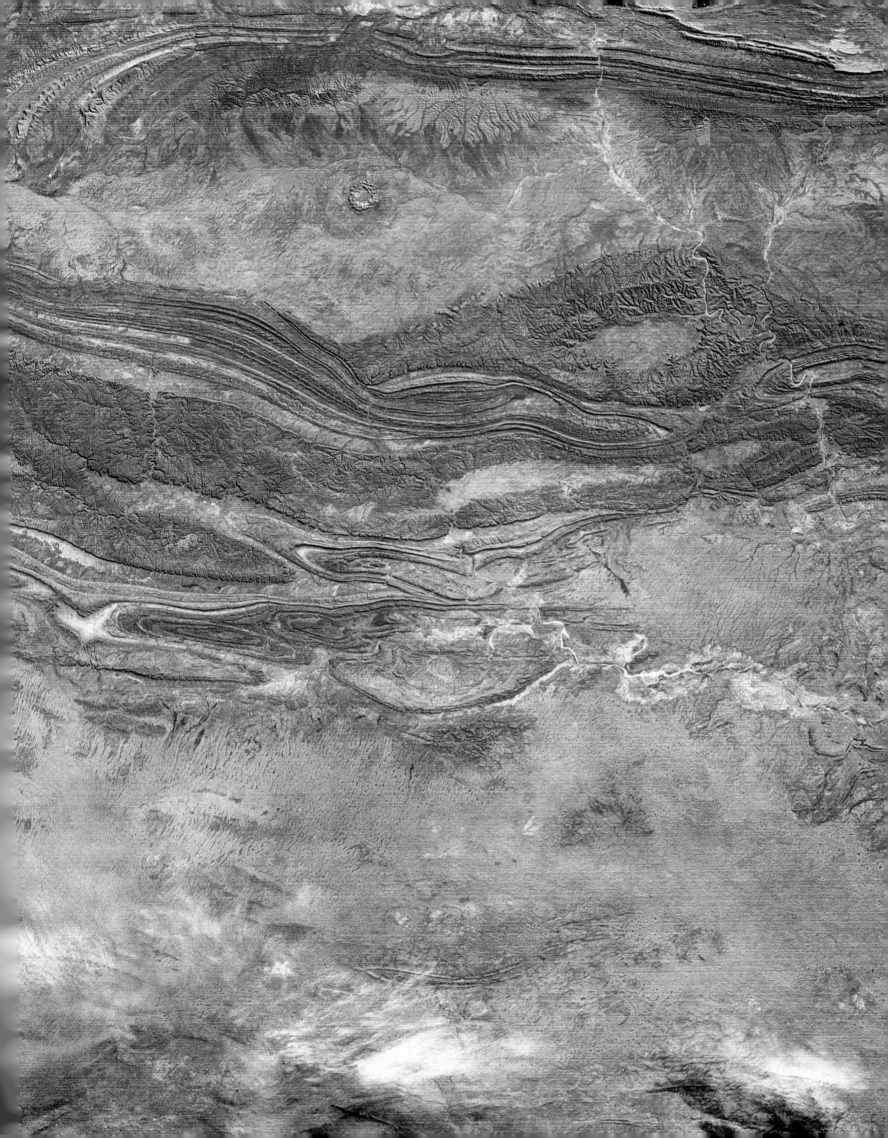

splendid example of an old mountain range. The Australians rather un-imaginatively refer to the centre of their vast island as the Centre, or sometimes the Red Centre, the latter bit of embroidery in acknowledgement of the pervasive red of the dusty landscape. Right at the centre of the Centre are the Macdonnell Ranges, which form the only real hills between the west and east coasts (Fig. 47).

The folded structure of these mountains shows up plainly in the image. Ancient sedimentary rocks were folded into anti-clines some 300 million years ago, and erosion later sliced horizontally through the folds, leaving their grain exposed like freshly planed pine. Although their structure was defined long ago in geologi-cal time, the ranges as we see them today are probably the result of relatively recent uplift. The Finke and Palmer rivers (top right and centre) appear to cut right across the folded strata, rather than being deflected into courses parallel with the ridges. This suggests that the rivers existed *before* the ranges were elevated to form hills, and that the rivers simply kept on carving downwards as the hills rose, preserving their original courses.

Mt Zeil, the highest point in the Macdonnell Ranges, reaches 1510 m (4954 ft). The height of the ranges yields some extra rainfall, but even so this is very arid country and the rivers are only seasonal. Permanent water-holes exist in the river-beds in a few places, and provide foci for small settlements. By far the best known of these is Alice Springs, whose world-wide fame far belies the population of only about 16,500. Green vegetation is confined to a few river-beds, notably the Palmer (right centre). Shortage of water naturally limits the development of the area, but the Macdonnell Ranges are becoming very popular with Australian tourists, who leave their crowded, cosmopolitan coastal cities to sample the realities of the 'outback' that is part of their cultural heritage, but which is alien to the everyday experience of the city-dwelling majority of Australians. They are rewarded with magnificent, but not grand scenery; stark mountains in a pure, unspoiled desert environment.

At top centre of the image is a remarkable structure that is quite unre-lated to the rest. It is the Gosse's Bluff Crater, formed where a small asteroid collided with the Earth. Early in its history, the Earth experienced a massive asteroid bombardment and its surface was heavily cratered, much as the Moon's is today. After the initial heavy bombardment, there was a dramatic decrease in the number of asteroid impacts. Since then normal pro-cesses of erosion have erased all trace of

those early collisions and only a handful of much younger impact craters is known to exist. Gosse's Bluff is one of that small number. The crater itself was originally about 20 km (12 miles) in diameter; what the image shows most clearly is a circle of sandstone about 3 km (2 miles) across that was uplifted to form a central ring in the crater.

Gosse's Bluff is one of the few places on Earth where geologists can study the results of impact cratering, perhaps the most important physical process that has shaped the surfaces of our neighbouring, less active planets. Lake Manicouagan in Quebec, Canada, provides a further, even more impressive example, captured on film in a remarkable photograph by the Shuttle 9 astronauts (Fig. 48). (Part of the Shuttle tail is visible in the lower right of the image.)

Manicouagan is one of the largest-known terrestrial impact craters, with a diameter of 65 km (40 miles), a mere dimple compared with the largest lunar structures which are more than 600 km (375 miles) across. It is believed to have

48 A remarkable image of a remarkable structure: the Manicouagan impact crater, Quebec, Canada. The Shuttle tail is visible at bottom right (Shuttle 9, natural colour). Image scale 1 cm = 5 km; map scale 1 cm = 70 km.

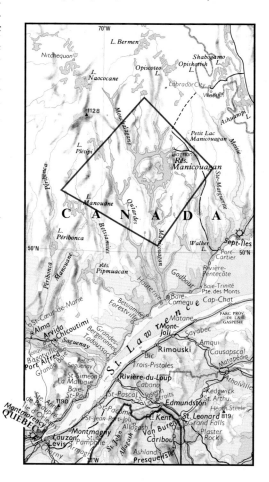

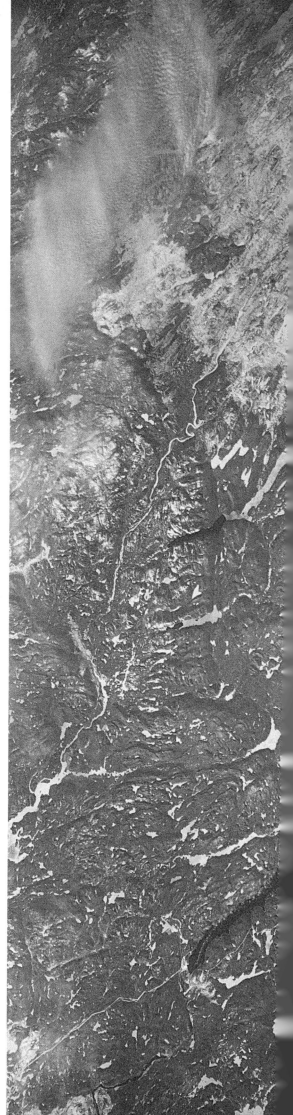

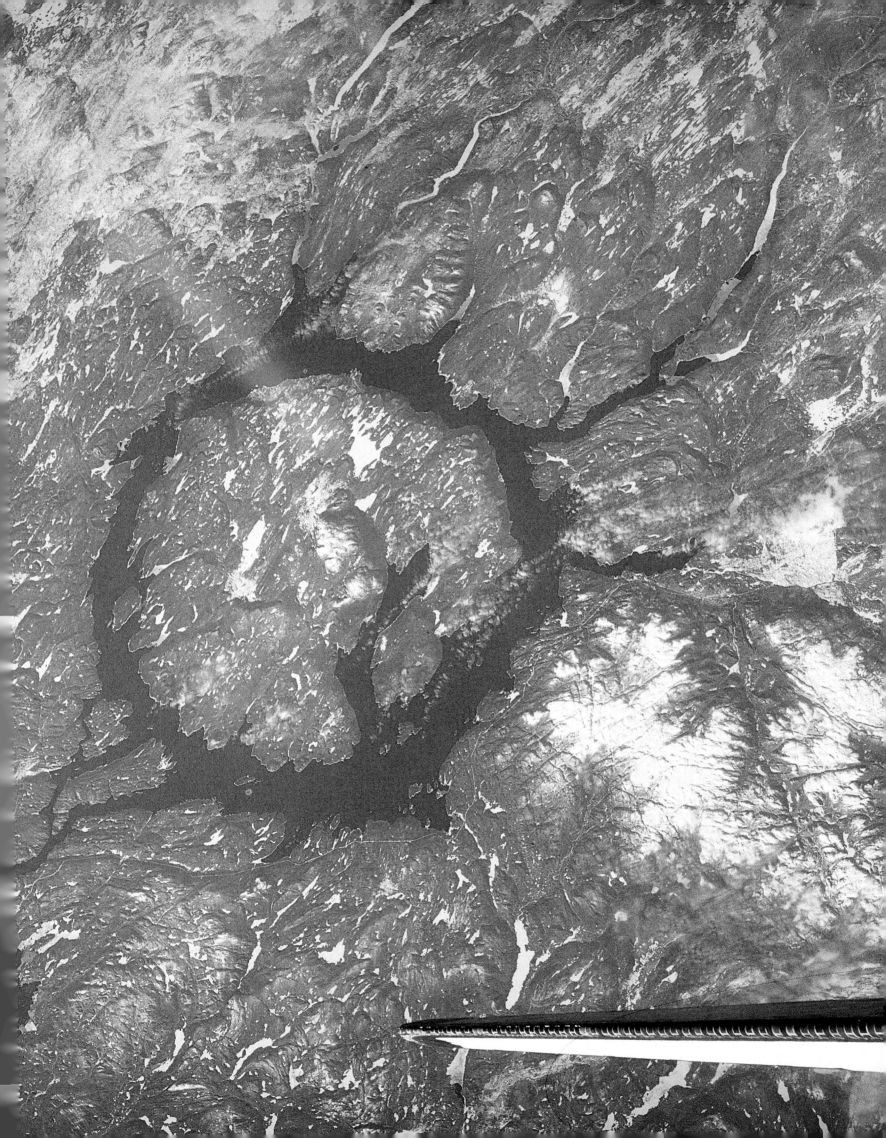

been formed some 200 million years ago with the impact of a much larger asteroid than that which excavated Gosse's Bluff. Large impacts produce much more complex structures than small ones. Again, their most obvious characteristic is their circular symmetry, but the shock waves produced on impact throw up a series of concentric rings, a little like the spreading circular waves set up when a pebble is thrown into still water. At Manicouagan, the concentric structure is beautifully displayed and emphasized in the perfect ring of the lake.

The lake itself is partly artificial, since it is actually a reservoir, ponded up by the building of the Daniel Johnson Dam across the Manicouagan river. In the lower left corner of the image, the dam can just be seen with a snow-covered road snaking across it and winding up to the reservoir. Although a 200-million-year-old structure might seem to be so ancient as to be irrelevant, impacts and impact structures have played a crucial role in the evolution of the Earth, a role that has only begun to be appreciated in the last few years. It is now realized that the catastrophic extinctions that took place 60 million years ago, when entire animal groups such as the dinosaurs and ammonites were abruptly and simultaneously snuffed out, were caused by the impact of an asteroid. The dust cloud raised by the collision veiled the Sun, causing permanent darkness and lowering temperatures dramatically. This impact may have involved an asteroid not much bigger than that which excavated Manicouagan crater.

Geologists did not really become aware of the importance of impact structures in the history of the Earth until the exploration of the Moon focused attention on them. Previously, the general philosophy had been that the Earth is moulded by slow, gradual changes rather than by catastrophic instantaneous ones. One consequence of the dawning awareness of the importance of impact structures was that a world-wide search was started for them, and many features with circular outlines were seized on as having originated from an impact. Many other kinds of geological phenomena can give rise to circular structures, however. Figure 49 illustrates just one of them, an intrusion of magmatic rock into the Earth's crust.

Situated some 80 km (50 miles) from the desolate Skeleton Coast of Namibia near Cape Cross (bottom left), the Brandberg Mountain (2606 m, 8550 ft) is an isolated mountain massif that rises far higher than any other point for hundreds of kilometres around. In prehistoric times it was a focus for the bushmen of the desert, whose paintings can still be seen on the walls of

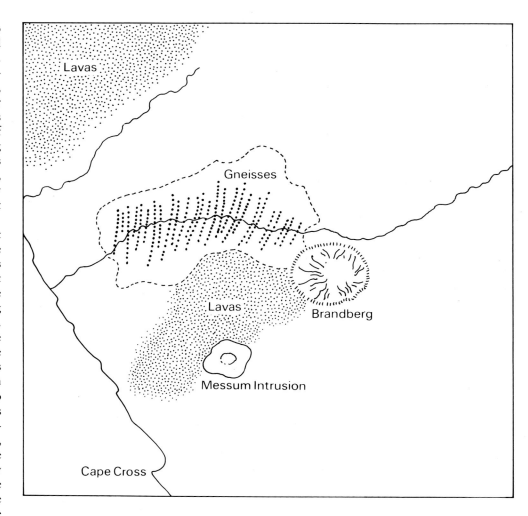

caves in the mountains. The massif is composed of a single mass of granite which rose bubble-like through the Earth's crust some 120 million years ago, shouldering aside the surrounding rocks as it did so. These rocks can be seen clearly in the image, encircling the margin of the intrusion, and tilted sharply upwards. Some of the rocks affected are ancient gneisses, distinguishable on the image by their grey tones and distinct lineated texture, most conspicuous along the course of the dry river valley at the centre. Immediately to the left of the Brandberg is an area of slightly rusty-red rocks; these are volcanic lavas belonging to what is known as the Karoo formation and are part of the same vast volcanic province as the lavas making up the Drakensberg plateau (Fig. 1). Slightly below the rusty-red rocks is a circular depression, ringed with concentric ridges, and looking rather like an impact crater. This is a second body of intrusive rocks, known as the Messum intrusion, much more easily eroded than the Brandberg, but part of the same episode of activity.

Clearly, eruption of the Karoo volcanic rocks and intrusion of related rock bodies such as the Brandberg over much of southern Africa was no trivial event. In fact, it is now believed that this extraor-

dinarily widespread episode of geological unrest was the forerunner of an even more important event in the Earth's history, the separation of South America from Africa and the formation of the Atlantic Ocean. For a long time before the actual rupture occurred, volcanic activity raged as the Earth's crust was softened up and fractured. Interestingly enough, the existence of a set of lavas in South America which matched the Karoo formation was used by some geologists as powerful evidence for continental drift for many years, but this evidence, along with many other clues, was rejected until relatively recently with the advent of the theory of plate tectonics.

According to this theory, the Earth's crust is divided up into seven major plates and a number of smaller ones, which are slowly shifting around relative to one another. Where plates move apart, oceans and new oceanic crust are created, and the line of separation is characterized by central volcanic ridges. In the Atlantic this ridge is mostly submerged, breaking surface only occasionally in volcanoes such as Tristan da Cunha and St Helena. Where plates converge, they can do so in two quite different ways. Where two continental areas come together, as in the Alps or Himalayas, great mountain belts are raised, but these typically lack volcanic

49 The Brandberg, Namibia: a circular intrusion that rose bubble-like through the Earth's crust (Shuttle 7, natural colour). Image scale 1 cm = 8 km; map scale 1 cm = 80 km.

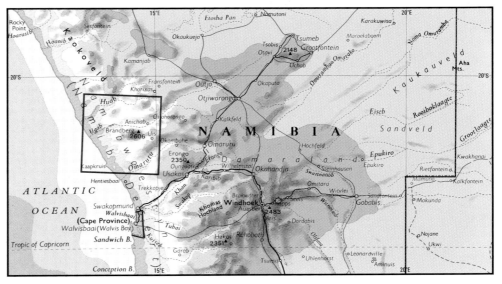

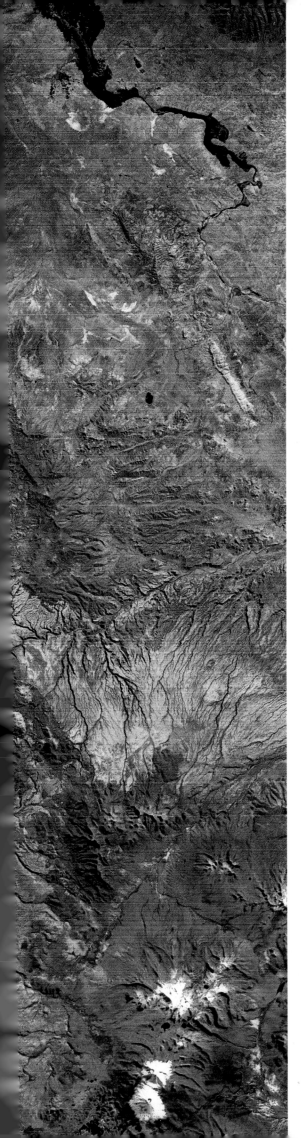

activity. Where plate made up of ocean crust converges on a continent, however, mountains are again thrown up, but these are characterized by chains of highly explosive volcanoes. Of these, the Andes provide much the best example.

Stretching over 8000 km (5000 miles) from the balmy Caribbean to the frigid waters of Magellan's Strait, the Andes are one of the Earth's greatest physical features. For much of their length, this range is dominated by soaring, 6000-m (20,000-ft) volcanoes. At both the northern and southern ends of the range, the volcanoes' inner fires are dangerously sheathed by ice, but in the centre of their length the Andes rise above the sands of the Atacama desert, the driest desert in the world, and their western slopes are barren and desolate. Most high mountain ranges are ornamented and softened by lakes and fertile valleys, but the central Andes are stark and severe. Here the great volcanoes encircle arid plateaux from which the intense tropical sun evaporates all moisture. Where one might have hoped to find lakes, there are searing salt-flats. But although the landscape is as harsh and hostile as any on Earth, it is hauntingly beautiful and has its own compensations for the visitor, such as the sight of haughty flamingoes stalking the shores of the salt lakes. Figure 50 shows part of the area of the Altiplano where three countries (Peru, Bolivia and Chile) come together. Although it has an average elevation of nearly 4000 m (13,000 ft), this part of the Altiplano is one of the least rugged sections of the central Andes.

Perhaps the most striking feature of this image is that every one of the peaks within it, without exception, is a volcano, and all but a small fraction of the rocks exposed are volcanic. Most of the volcanoes are rather old – a few millions of years – and extinct. They form the clusters of symmetrical, flower-like patterns that dominate the upper left of the image, most of which are in Peru. The deeply incised, radial gullies that give these volcanoes their distinctive shape cut right into the heart of the cones, clear evidence that the volcanoes are truly extinct and not merely dormant. Originally, these extinct cones probably rose to 6000 m (20,000 ft), but now most have been eroded down to less than 5000 m (16,400 ft). In the lower right-hand corner there are some active volcanoes, part of the Nevados de Payachata

50 Rosette-like shapes of eroded volcanoes stud the Andean Altiplano in the frontier region where Bolivia, Peru and Chile meet (Landsat, false colour). Image scale 1 cm = 5.5 km. See p. 110 for location map.

range. These exceed 6000 m, have small but distinct summit craters, and are flanked by young lava flows. The deeply gullied, light-toned areas at right centre consist of huge expanses of ash deposits, accumulated from successive eruptions of the wealth of surrounding volcanoes.

To a geologist, the rosette-like patterns made by the dissected volcanoes convey an immediate message. The 'petals' defining the rosettes are the results of glacial erosion, which scooped out deep 'U'-shaped valleys radially around each volcano. Today, of course, the glaciers are long gone, but the evidence of their existence is quite unequivocal. The timing of ice movements is not nearly as well known in the southern hemisphere as in the northern, but it is possible that central Andean volcanoes were draped with ice as little as 10,000 years ago. At the present day, there is very little water flowing through the great canyons (*quebradas*) that are so conspicuous in the bottom left corner of the image. Although some of these exceed the Grand Canyon in scale, they are not so spectacular because their sides slope more gradually. The climate on the western slopes of the Andes is now exceedingly arid and it seems that the great canyons could only have been gouged out during the Ice Age, when copious amounts of melt-water flowed from ice-covered mountains.

None of the rivers on the eastern slopes find their way to the sea, but instead flow no further than the Altiplano. The Mauri, at right centre of the image, cuts a deep gorge through thick volcanic ash deposits (pale tones) before winding into Lake Titicaca (top right). Slightly further south, where the climate is drier, much more evaporation takes place, and lakes are either seasonal or completely dried-up, as Figure 51 shows. This Shuttle picture displays lakes of two different kinds. At the bottom, the glistening white expanse is the surface of the Salar de Uyuni, the largest salt-flat in the world, over 100 km (62 miles) across and perhaps the only place on Earth from which one can see the Sun both rise above and set below an unbroken horizon of salt.

At the top left of the image, the extraordinarily bright green waters of Lake Poopo are prominent. Although this lake is very shallow, it never dries out completely because it is constantly replenished by the overflow from Lake Titicaca, to which it is connected by the southwards-flowing Rio Desaguadero. Lake Poopo has been little studied, and its green colour is not fully understood. It may be due to the growth of a species of freshwater algae that can thrive in the shallow, brackish water. It has been

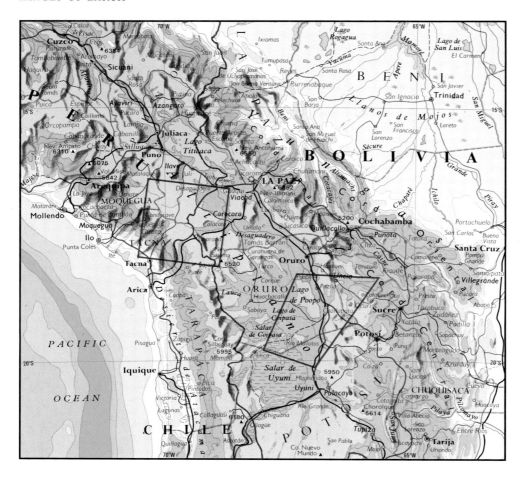

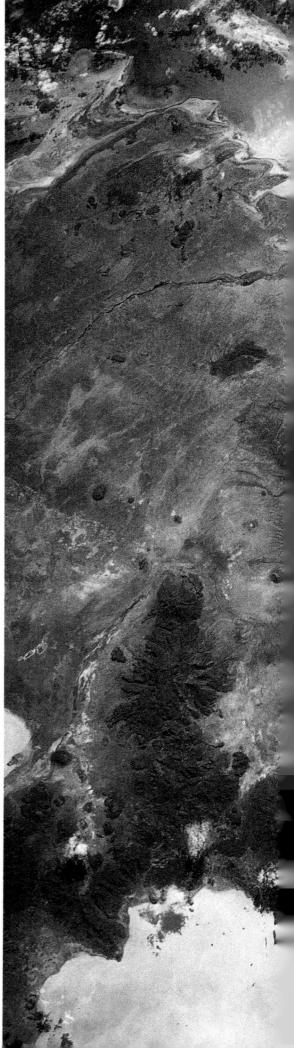

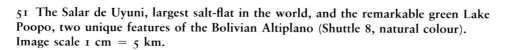
The map shows the area of Figure 50 (left) and Figure 51 (right). Map scale 1 cm = 80 km.

51 The Salar de Uyuni, largest salt-flat in the world, and the remarkable green Lake Poopo, two unique features of the Bolivian Altiplano (Shuttle 8, natural colour). Image scale 1 cm = 5 km.

52 One of the Earth's largest volcanic
structures, the Cerro Galan caldera,
north-west Argentina, first discovered on
spacecraft imagery (Landsat, false
colour: the same structure is shown in a
natural-colour photographic view in
Figure 42, centre right). Image scale
1 cm = 4 km; map scale 1 cm = 80 km.

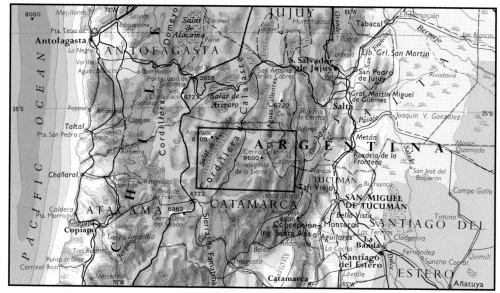

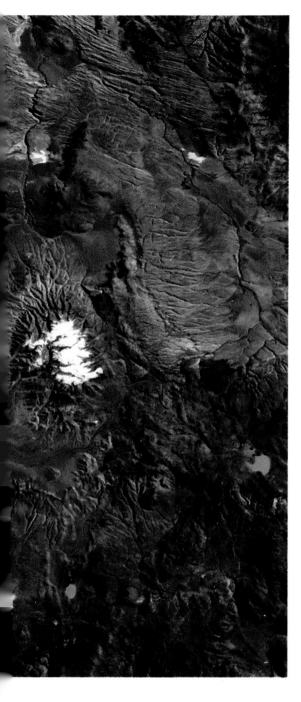

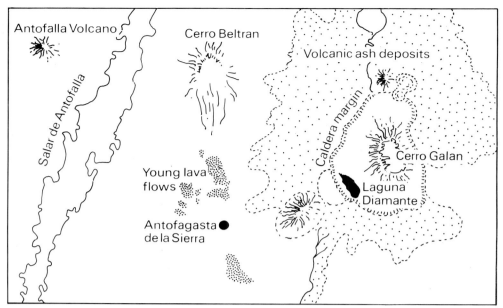

previously, and complete air photograph coverage existed, the caldera had not been detected because it was camouflaged by its own sheer scale. On the ground, inside the walls of the caldera, its enclosed shape is not immediately obvious – one has to know it is there to appreciate it.

The caldera was formed by a stupendous eruption about 2.5 million years ago and is one of the Earth's largest volcanic structures. The eruption ejected at least 1000 cubic km of pumiceous ash, which is now exposed in the greyish, deeply-gullied terrain around the northern parts of the caldera. So much material was erupted from the subterranean magma chamber that its roof collapsed downwards, leaving a vast subsidence crater at the surface which may have had walls more than 2000 m (6500 ft) high. This depression was probably soon occupied by a lake. Later, the caldera floor was heaved up once more to form the prominent central snow-covered peak of Cerro Galan, about 6600 m (21,600 ft) high, one of the loftiest mountains in northern Argentina. Upheaval of the caldera floor caused most of the water in it to be displaced; all that now remains is the deep blue lake in the southwest of the caldera, Laguna Diamante. Since the major eruption, minor volcanic activity has led to the extrusion of lava flows on the northern part of the caldera rim, and also in the floor of the valley west of the caldera, where some very dark lava flows are conspicuous. Within the caldera, hot springs still gush large volumes of boiling water to the surface.

If it were located in a populated area, the caldera could prove to be a major geothermal resource. As it is, the only inhabitants in the area live in the small hamlet of Antofagasta de la Sierra, located in the vegetated area (red) on the west of

the image. This region is so high and arid that minimal subsistence farming is all that is possible. The floor of the caldera is at an average height of 4800 m (15,700 ft) and for the most part entirely barren. Around the shores of Laguna Diamante, however, there is grazing for hundreds of llamas, and flamingoes and a range of other birds flourish on the salty waters.

Massive eruptions like that which formed the Cerro Galan caldera inject so much dust into the Earth's atmosphere that their effects are profound and worldwide. Although the Cerro Galan eruption was one of the largest known to man, it took place so long ago that we do not know exactly what effect it had, but the consequences of eruptions that have taken place within well-documented history provide fascinating parallels. The largest eruption for which we have detailed records was that of Tambora, on the Indonesian island of Sumbawa in 1815 (Fig. 53).

Sumbawa is one of the hundreds of islands, large and small, that form the country of Indonesia. The island chain extends over 4500 km (2800 miles) from west to east, and is dotted with active volcanoes along its entire length. Several ancient, eroded cones are visible on the left side of the image, but the site of the 1815 eruption was the distinct circular crater at top right. Now 2821 m (9256 ft) high, Tambora volcano is believed to have once stood well over 3000 m (9800 ft), but it was decapitated during the eruption and an 8-km (5-mile) diameter crater now occupies the site of the conical summit of the original volcano. It has been estimated that over 50 cubic km of ash was ejected from the crater, two and a half times as much as during the much better known eruption of Krakatau in 1883, and a hundred times as

suggested that the lake might be fed by hot springs that contribute both warmth and mineral nourishment, but this is uncertain.

A traveller on the Altiplano of Peru or Bolivia would have little difficulty in recognizing the conical shapes of the many volcanoes there. But many larger volcanic structures exist in the central Andes, some of them so large that their shape can only be appreciated from space. Some important discoveries have been made since space images have become available. The enormous volcanic caldera in Figure 52 was first noted on photographs taken by astronauts aboard Skylab 4 in 1974. Located in a remote part of northwest Argentina, the Cerro Galan caldera is 35 km (22 miles) in longest diameter. Although geologists had visited the area

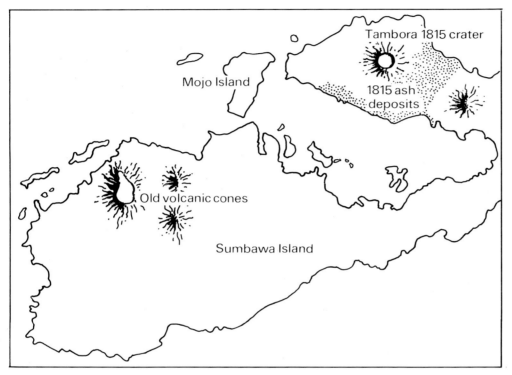

Tambora 1815 crater

Mojo Island

1815 ash
deposits

Old volcanic cones

Sumbawa Island

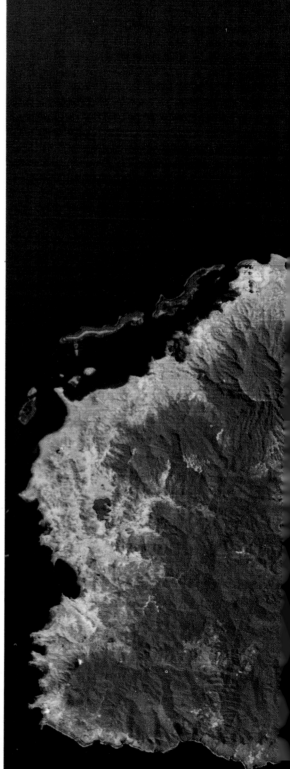

53 The Tambora volcano, Sumbawa Island, Indonesia, scene of the world's most violent historic volcanic eruption (Landsat, false colour). Image scale 1 cm = 6 km; map scale 1 cm = 70 km.

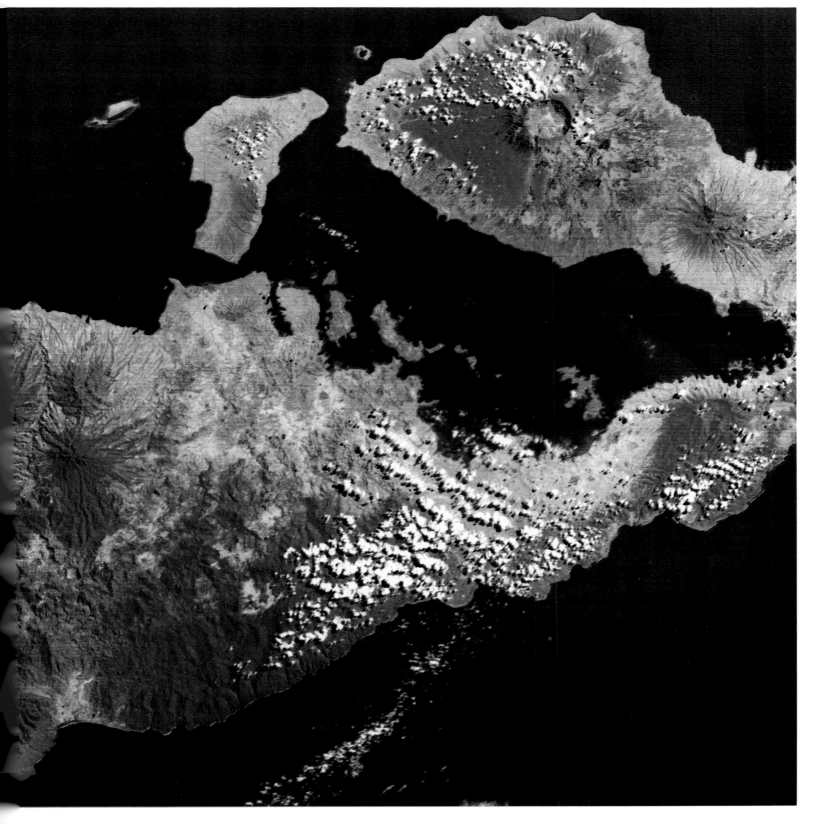

54 A tranquil mantle of snow ornaments
some destructive volcanoes in the
southern Kamchatka peninsula, USSR
(Shuttle 9, natural colour). Image scale
1 cm = 4 km.

The map (below right) shows the
location of Figure 54 (above) and Figure
55 (below). Map scale 1 cm = 200 km.

much as the dramatic and highly publicized eruption of Mt St Helens in the USA in 1980.

Ash fell over a huge area, causing day to turn to night in Java, 500 km (300 miles) distant. Ships at sea reported masses of floating pumice more than four years later. So much ash and volcanic gas entered the atmosphere that a pall spread round the world, cutting down the light from the Sun, and affecting the global climate. So severe were the effects of this pall that crops failed in both North America and Europe the following year (1816), which became known as 'the year without a summer'. There were frosts in New England in July and August, and in Wales it was reported that there were only three or four days without rain between May and October. Apart from its gloomy effects on global weather, the eruption also had an unexpected benefit: the haze of dust and ash produced most spectacular sunsets for six months after the eruption. It has been suggested that these glorious twilight hues inspired some of the best works of one of Britain's greatest artists, J. M. W. Turner, the 'painter of light'.

More recently, the Tambora eruption has been providing a data base for a more sinister application. It has been realized that a major nuclear exchange of 5000 megatons could raise such a heavy veil of smoke that sunlight would be entirely blocked over most of the Earth, and that a catastrophic 'nuclear winter' would result. Since there is little if any information bearing directly on this issue, the scientists involved are looking closely at the Tambora eruption and its effects in order to establish a firm baseline from which to estimate what might happen if eventually deterrence fails to deter.

The tropical coral-fringed volcanic islands of Indonesia and the high, dry calderas of the Andes lie on opposite sides of the Pacific Ocean, but they form part of the same great volcanic belt, the so called circum-Pacific Ring of Fire, which loops erratically around the entire Pacific in a series of island arcs and mountain belts. Figure 54 shows another aspect of the Ring of Fire, this time on the southern tip of the Kamchatka peninsula, immediately south of the major city of Petropavlovsk.

Kamchatka is of great strategic significance to the Soviets, giving the USSR access to much of the west Pacific. Petropavlovsk is a major naval port, where nuclear submarines are based – jet fighters were scrambled to shoot down the Korean airliner in 1983 after it had flown directly over this highly sensitive area. Because of its military importance, coupled with the

Soviet's customary paranoia concerning state security, it is difficult to learn much about the volcanoes of Kamchatka. Even scholarly papers on the geology of the area are published with only the sketchiest of maps, devoid of normal cartographic information such as contours, roads, map scale or north point. Scientifically, it is rather sad, since Kamchatka is one of the most active volcanic provinces in the world, containing at least twenty volcanoes that are known to have erupted in recent times.

Two of these active volcanoes are visible in the image, their profiles mantled with snow and crisply defined by the low angle of the sun, which beautifully reveals the contours of the land. At top centre is the cone of the Gorelyy Krehbet (1829 m, 6000 ft), with an elongate series of craters located at its summit. This volcano, which last erupted in 1931, is itself located at the centre of an older caldera, about 9 km (5½ miles) in diameter, which completely encircles it. Ash-flows from the massive eruption that created the caldera are visible to the northwest. They are covered by snow, but conspicuous because of the deep gullying.

Directly south of Gorelyy Krehbet is the second volcano, Mutnovskaya (2323 m, 7621 ft), a complex cone, with the youngest craters on its northwest side. This volcano is believed to have last erupted in 1945, but there is little in the image to suggest that this eruption was a large one. At the left centre of the image is another large volcano, partly obscured by thin cloud gathering near its summit. This structure appears to be rather old and eroded, however, and there are no records of its having erupted in historical times. At top right, yet another cone is present, casting a perfect pointed shadow of its symmetrical profile on the wall of the valley opposite. This volcano also has no record of recent activity.

With such an abundance of volcanoes, it is natural that Kamchatka should be rich in geothermal energy reserves. It was indeed on the Kamchatka peninsula that Soviet engineers first harnessed geothermal waters for electric power and heating. As the snow-covered mountains lapped by the frigid waters of the north Pacific suggest, the winter climate of Kamchatka is severe and the inhabitants need all the heat they can get. It is not only the human inhabitants that benefit however. In one or two localities in Kamchatka, entire plant and animal communities survive the winter cosily in narrow valleys in which temperatures are maintained many degrees above their surroundings by numerous hot springs and geysers. These exotic habitats are of considerable interest to naturalists,

55 Tyatya volcano, Kunashir Island, in the Kurils, USSR, in vigorous eruption (Landsat, false colour). Image scale 1 cm = 8.7 km. See p. 117 for location map.

but few, even from the USSR, ever see them.

From Kamchatka, the Ring of Fire extends southwards for nearly 1000 km (625 miles) along the great island arc of the Kurils, where some 33 active volcanoes are scattered along the 30 large and 20 small islands and innumerable reefs and rocks. Politically, these remote outliers of the USSR have had a chequered history. Originally Russian, they were ceded by Russia to Japan in 1875 in exchange for Sakhalin Island, which lies close to the Russian mainland. Although the southernmost island in the Kurils is less than 50 km (31 miles) from the coast of Hokkaido, Japan's northernmost island, the Kurils were returned to the USSR after World War II in order to deprive the Japanese of the immense strategic advantages that the island chain offered. Whatever their significance in the geopolitical chess game, the Kurils are not particularly attractive for human habitation. Humidity is high all year round, with cold winters and hot, foggy summers. Apart from a little sulphur mining, there is not much of economic interest in the islands, where the main activities are centred around fishing, crabbing, sealing and whaling.

Whatever the islands lack in postcard charms, they make up for in terms of

volcanological interest. One of the most remarkable satellite images ever acquired was recorded by the first Landsat satellite in July 1972; it shows a towering plume of ash and dust rising from an eruption of the Tyatya volcano, at the northern end of Kunashir Island. Tyatya had been peaceful for over 161 years since its last eruption in 1812 when it suddenly burst into life two days before the image was acquired. Violent explosions took place, audible in the tiny fishing villages on the island up to 80 km (50 miles) distant from the volcano, and the vast eruption plume climbed several kilometres into the air.

Careful inspection of Figure 55 shows that the plume is rooted on the southern edge of the parent volcano, defined by a whitish ring. Soviet scientists reported that a new vent had opened on the south flanks of the volcano, and that all the activity was focused there. Dark ash fell over large areas, accumulating up to nearly 1 m (3 ft) in thickness near the volcano, and killing the dense scrub that previously clothed it. The fishing village of Tyatino, only a few kilometres from the active vent and directly beneath the eruption plume, was severely damaged. When the image was acquired, the wind was carrying the plume away to the southeast, but it had previously been much more widespread.

Dark ash covers the northeastern tip of the island so completely that the coastline is barely distinguishable, but some small smoke plumes from burning scrub can just be made out. Towards the west, the margins of the ash-fall are quite distinct, the unburned vegetation showing up in bright red.

Although many other eruptions have taken place since that of Tyatya, few have been so clearly imaged. Often, of course, the volcano is obscured by clouds, or else the eruption cloud itself is so extensive that little detail can be seen in the ascending plume. Here, however, the wind was sufficiently strong that the plume is extended downwind, but it is not so strong that the texture in the eruption column has been smeared out. Upwind of the volcano, another impressive, but quite coincidental effect can be seen: a sort of bow-wave of cloud, caused by the 1822-m (5978-ft) volcano jutting into the air-stream above it. Just as water in front of a large ship's bow is piled up into waves well ahead of the ship, so the wind pattern can be disturbed well ahead of any mountain in its path. The wavelike rise and fall of the air masses in the bow-wave causes moist air to condense to ripples of cloud which are clearly visible in the image, but fortunately these had dissipated in the region of the volcano itself. Exactly the same wave patterns can be seen near the volcanoes to both north and south, one of them around the Rikorda volcano on Iturup Island (top right), and the other on the long peninsula jutting out from Hokkaido's north shore (bottom left).

Volcanic activity is the only geological process that can increase the height of a mountain at a rate that is perceptible on a human timescale. Vesuvius, for example, has added several hundred metres to its height since its catastrophic eruption in AD 79. Every other geological process that takes place at a directly measurable rate involves the weathering away of mountains and hills in a slow but steady lowering of the topography. Erosion by streams and rivers is a matter of everyday experience, and their effects on landscapes are visible in almost every corner of the world. Many of the most spectacular landscapes, however, the sort that people are prepared to travel long distances and pay large sums to see, have been shaped by glaciers. This is true of all the soaring peaks in all the great mountain ranges of the world.

Of all the magnificent possibilities around the world that could have been chosen to illustrate the action of glaciers, one of the most remote and least known has provided a particularly sumptuous and informative image: Bylot Island, in the far north of Canada (Fig. 56). This island, whose shape is curiously reminiscent of the Isle of Wight, is separated from the much larger Baffin Island by the narrow water of Eclipse Sound (right) and by the somewhat improbably named Navy Board Inlet (left). Snow mantles the crests of the Byam Martin Mountains that form the backbone of the island, and which reach 1800 m (5900 ft). Not surprisingly in view of its high latitude (73°N), the whole area of the image is effectively uninhabited, though there is a tiny Eskimo settlement at Pond Inlet (lower right, but not discernible).

The image was acquired during the height of the northern summer, in August 1974. Thus, all the previous winter's snow had been burned off by the sun, except on the highest peaks of the mountains, and the permanent ice of the valley glaciers is crisply exposed. Several major and dozens of minor glaciers are visible. They illustrate perfectly how many mountainous areas of northern Europe, such as Scotland, which are now comparatively mild and ice-free, would have looked a mere ten or twenty thousand years ago, when their topography was being sculpted by the great glaciers of the last Ice Age. Bylot Island's glaciers display almost all the features typical of valley glaciers. For example, in grinding away at the valley walls that enclose it, each glacier picks up a burden of rock debris know as a marginal moraine and these moraines form stripes on either side parallel to its length. Where two glaciers merge, the marginal moraines combine to become a broad central moraine, visible as a dark stripe in the middle of the ice flow, such as that at top right. The merging of many tributary glaciers forms many such stripes, and these often become contorted as the glacier spreads out towards its snout (top centre).

At the snouts, rapid melting takes place, and the melt-water forms shallow braided rivers that carry a heavy load of the fine rock powder that the glacier has ground beneath it. Where these streams enter the sea, sediment deltas are built up and rock flour makes a distinct white plume against the indigo of the water. Where the glacier itself enters the sea, chunks of ice break off to form floating islands – icebergs – small examples of which can be seen at top left. In Eclipse Sound, there are ice islands of a different kind. These are ice-floes, chunks of the sheet of ice that would have covered the whole of the sound in winter, and which were breaking up rapidly when the image was taken, as the swirl of small floes at lower right shows.

To the north of the island, the deep water of Baffin Bay is, naturally enough, always cold. The land to the south, however, warms up quickly in summer, and this results in a characteristic weather pattern, with heavy fog and clouds persisting out over the sea, but becoming thinner and less persistent inland. While Eclipse Sound is often sunny and cloud-free, Bylot Island itself is not. We should be grateful that the skies were so clear in August 1974, making this impressive image possible.

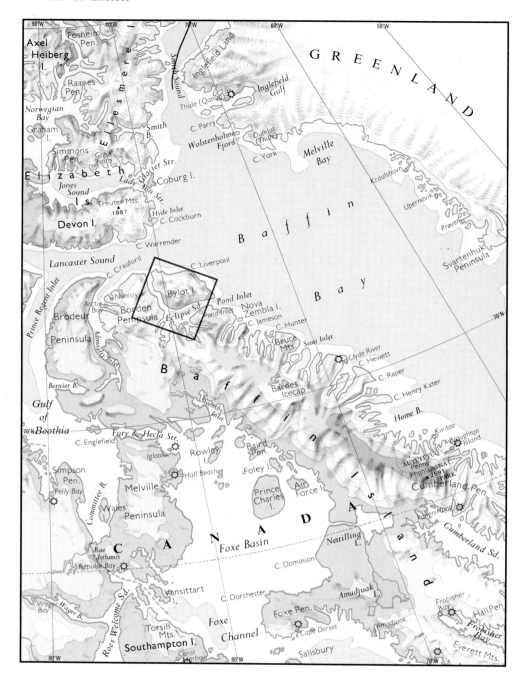

56 Finger-like glaciers gouge through the Byam Martin Mountains, Bylot Island, in the far north of Canada (Landsat, false colour). Image scale 1 cm = 5.5 km; map scaie 1 cm = 100 km.

6 The Fragile Veil

To most people who use satellite imagery, clouds are a distinct nuisance because they obscure what lies beneath. At any given moment, they cover 20 to 30 per cent of the Earth's surface. In tropical regions, particularly the more mountainous ones, cloud cover is so frequent that cloud-free images are virtually unobtainable. Furthermore, in many warm, humid regions, although the nights may be clear, clouds build up very rapidly in the morning and often cover much of the sky by the time that Landsat satellites come overhead at around 9.30 am. Clouds are such a problem that a variety of different radar-based techniques have been developed which can 'see' through the cloud to the ground beneath, both from aircraft (Side Looking Airborne Radar) and from space (Shuttle Imaging Radar).

From time to time, though, even the most philistine geologists have to stand back and admire the sublime beauty of cloud formations. Although meteorological satellites are numerically the most important single sources of images of Earth, ironically they do not provide good cloud pictures. This is because satellites that have been especially designed for meteorological work tend to show vast areas at a time and do not display individual cloud formations. It is only when clouds creep inadvertently on to the much more detailed images used by geologists, or are deliberately photographed by astronauts, that the glorious richness of cloud shapes and textures is revealed.

Cloud systems range in size from the gigantic spirals of hurricanes, which are often several hundred kilometres across, to the smallest puffs forming in currents of rising air above a hillside. Each has its own story to tell of the subtle interactions between air and water that produce the shifting weather patterns and cloudscapes that influence our daily lives and form a vital part of our visual impression of the world. What would the great landscape painters have done if there were no clouds?

From the earliest days of manned spaceflight, the endless variety of cloud patterns and structures has been a source of delight to astronauts, and the broader view that they get from their lofty orbits often helps to transform what would at the surface of the Earth seem to be a confusing complex of cirrus or cumulus into an elegant and intelligible cloud system. On the Apollo 7 mission in 1968, for example, some magnificent views of Hurricane Gladys (Fig. 57) were obtained from a height of 150 km (94 miles). Over the years, meteorologists had patiently worked out what goes on in a hurricane from countless surface and aircraft observations of temperature, pressure and wind direction, so the image contained no surprises, but it encapsulated all the previous work in a single breathtaking demonstration of the process.

How a hurricane is formed (northern hemisphere). Winds blowing in towards a low pressure area (depression) are deflected to the right by Coriolis forces, and generate an anticlockwise spiral circulation around the centre of the depression. Powerful vertical convection develops at the centre of the depression (the eye of the hurricane) with winds spiralling outwards (clockwise) at an upper level.

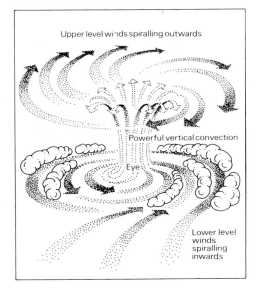

57 An elegant spiral seen from above, but a destructive maelstrom below. Hurricane Gladys over the Gulf of Mexico (Apollo 7, natural colour). Image scale 1 cm = 6 km at centre.

58 A last view of Earth. A decaying circular storm photographed off Australia by the Apollo 11 astronauts as they left Earth for the Moon (Apollo 11, natural colour). Image scale 1 cm = 15 km.

59 Off the Atlantic coast of Morocco, prevailing winds often set up eye-catching spiral eddies of cloud (Shuttle 9, looking southeast, natural colour). Image scale 1 cm = 15 km.

Hurricanes (or typhoons) will only form where there is both sea-water warmer than 27°C (80°F), and a tropical depression at low, but not equatorial latitudes. As air flows into the depression, a spiral motion develops as a result of forces set up by the Earth's rotation (Coriolis forces). The incoming air has been warmed and moistened by blowing over the sea and this air rises at the centre of the spiral. As it does so, the moisture in it condenses and liberates more heat and this in turn causes the air to rise even faster. Thus, a powerful rotating and convecting system is set up, driven by the energy of the warm sea, which can work itself up over a period of many days into a vicious and powerfully destructive storm. Hurricanes such as Gladys are born far out to sea, and then move slowly northwestwards (in the northern hemisphere). Eventually, they turn north or northeasterly and begin to run out of energy as they pass over cooler surface waters.

If a hurricane should cross a coastline, of course, the effects can be devastating, not only because of the powerful winds, but also because the winds and low atmospheric pressure may combine to whip up a storm surge in the sea, thus raising sea-level locally by several metres. Storm surges have been responsible for catastrophic inundations of many low-lying coastal areas, particularly in the Gulf of Mexico and the Bay of Bengal.

In one of the worst natural disasters ever recorded, a hurricane came ashore near Calcutta on 7 October 1737, causing a 12-metre (39-ft) storm surge which killed an estimated 300,000 people. Hundreds of thousands more died in November 1970 when another hurricane swept over this low-lying area of the Ganges delta, and countless thousands died in the numerous lesser storms that raged in the 230 years between the two great catastrophies.

Hurricane Gladys was photographed west of Naples, Florida, on 17 October 1968, while still an active system. The eye of the hurricane is sharply defined by the disc-like cloud mass at dead centre, nearly 20 km (12½ miles) in diameter and at a height of 15 km (9 miles) above the Earth. The air in the cloud mass has been pumped up from the surface of the ocean, cooling as it rose and spreading out to form the cloud disc in the image. In the spiralling arms of the system chains of towering thunderclouds can be seen, while to the left of the eye there is an expansive stratum of shapeless cloud. Beneath this formless layer, heavy rain is falling – all the world's rainfall records have been connected with the spiralling arms of hurricanes, where a metre often falls in under 24 hours.

When the picture was taken, maximum winds near the eye were 65 knots; two days later, the eye had drifted northwards towards Jacksonville, Florida, and winds reached 85 to 90 knots. Subsequently, the hurricane moved rapidly northwards and, although it was travelling over the cooler waters of the Atlantic, survived as far north as Nova Scotia, where it arrived on October 20 as an 'extra-tropical low'.

In 1969, Apollo 11 astronauts photographed a rather similar rotating cloud system west of Australia (Fig. 58), as they left the Earth for the first landing on the Moon. The spiralling clouds in this picture, however, did not amount to a hurricane, but were the swirling remnants of a dying storm of a sort that is fairly typical over the cool waters of the area. Although not in the same league as Gladys, the storm was initiated in much the same way, but lacked the powerful convecting centre that defines the eye in a hurricane. It also differed from Gladys in one very significant respect: the spiral was rotating clockwise, the opposite to the rotation it would have had in the northern hemisphere. North Americans and Europeans often jokingly suggest that bath-water swirls down the plughole differently in antipodean regions. In the case of revolving storms, this is true.

The technical term for a spiralling depression is a cyclone, but this general description covers wind patterns of very varied origin and on different scales. Figure 59 shows a splendid little circular cyclone off Agadir, Morocco. Although apparently so similar in form to a hurricane, this cyclone has a very different character and forms a consistent, almost stable wind pattern. Prevailing northeast winds sweep down past Cape Rhir, where the coastline bends sharply eastwards, with the strongest winds over the sea where their path is unobstructed. Because of the different wind velocities over land and sea, a sort of large-scale eddy is set up whenever the winds are strong enough and whenever there is enough moisture in the air to form clouds. As Figure 59 shows, the clouds are so closely packed that they almost form a continuous sheet, but they do not have much vertical development.

Some of the simplest and most elegant cloud patterns are those that develop where winds sweeping unhindered over the open ocean encounter solitary, mountainous islands. The mountain peaks stick up into the otherwise smooth, uniform airflow and disrupt it, producing cloud effects that extend for hundreds of kilometres downwind. From the island itself or the sea surrounding it the movements of the air currents are hard to make out, but clouds on satellite images trace them unambiguously.

Pulling an oar through calm water creates two rows of swirling vortices, with each alternate vortex rotating in opposite directions. Known as von Karman vortices after the brilliant Hungarian scientist who first studied them, they are beautifully displayed in a photograph taken by Skylab 3 astronauts of the wind's wake behind Guadalupe Island, off Baja California, Mexico, in 1973 (Fig. 60). Here, the hilly island has the same effect on the northerly wind as an oar in water, only on a much larger scale, creating swirls up to 50 km (31 miles) in diameter which extend downwind for more than 200 kilometres (125 miles). Although small-scale laboratory experiments show that pairs of spirals with opposite rotations should be produced, only the anticlockwise (cyclonic) spirals are well developed in the image. A hint of a clockwise swirl can be seen immediately south of the first large anticyclonic vortex, but they are probably being suppressed by the forces which favour anticlockwise motion in the northern hemisphere.

In detail, the spiralling anticlockwise pattern is constantly shifting and changing, but interestingly enough, the pattern of northerly winds blowing over Guadalupe Island is sufficiently consistent for von Karman vortices to be regularly developed, and astronauts have come to recognize them as a kind of orbital landmark. Ten years after the Skylab mission, the Shuttle 7 astronauts took a photograph showing the same street of vortices off Guadalupe (Fig. 61), here becoming wider and more diffuse downwind. Again, only anticlockwise spirals are developed.

Meteorologists are naturally delighted to have pictures of von Karman vortices, but an understanding of how such vortices form is of first importance in many other fields – any activity, in fact, that involves moving fluids, either liquid or gas. In 1940, for example, the Tacoma Narrows Bridge in the state of Washington, USA, collapsed spectacularly only four months after its completion, in a wind of only 70 km (44 miles) per hour, causing a major scandal and embarrassment to the engineering profession. It was von Karman himself who showed that, among other factors, turbulent vortices set up by the bridge in the wind caused such violent oscillations and twisting of the structure that they rapidly resulted in its failure. Bridge design has never been the same since. Modern suspension bridges are designed and finely tuned on the basis of the same aerodynamic principles that are used in the design of aircraft, contrasting sharply with the massive, heavy structures of the early part of the century, when aerodynamics were ignored and safety was obtained by brute strength and rigidity.

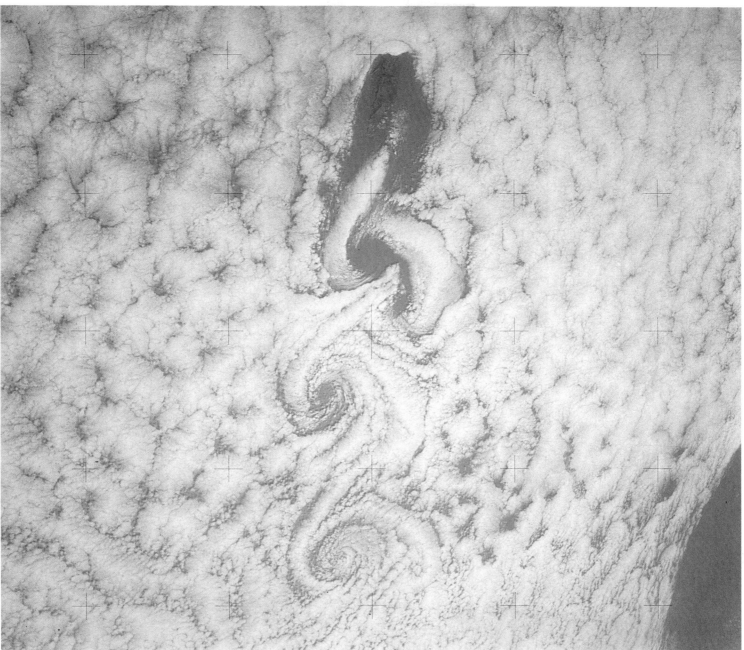

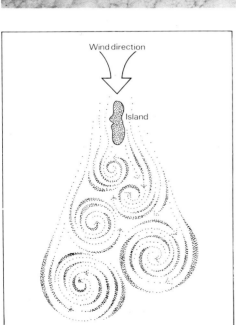

60 The wind's wake: spiral vortices set up downwind of Guadalupe Island, off the coast of Mexico (Skylab 3, natural colour). Image scale 1 cm = 15 km. The diagram shows an idealized pattern of von Karman vortices downwind of an oceanic island.

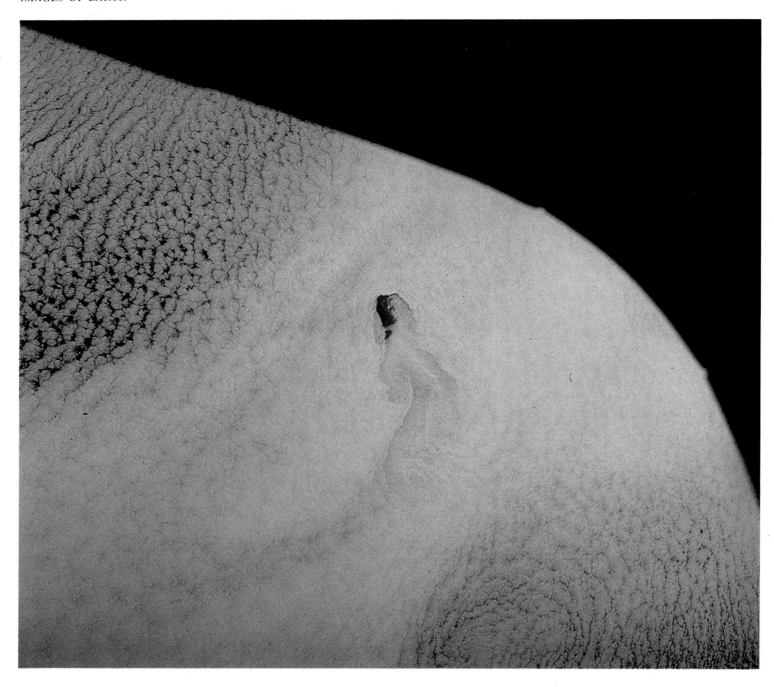

61 The wind's wake, ten years on. Wind
patterns around Guadalupe Island are so
consistent that these vortices
photographed on the seventh Shuttle
mission in 1983 are almost identical to
those in Figure 60 (Shuttle 7, natural
colour). Image scale 1 cm = 20 km.

62 A train of fair-weather cloud streams
200 km (125 miles) downwind in a long
thread from Socotra Island in the Indian
Ocean (Shuttle 1, natural colour). Image
scale 1 cm = 10 km.

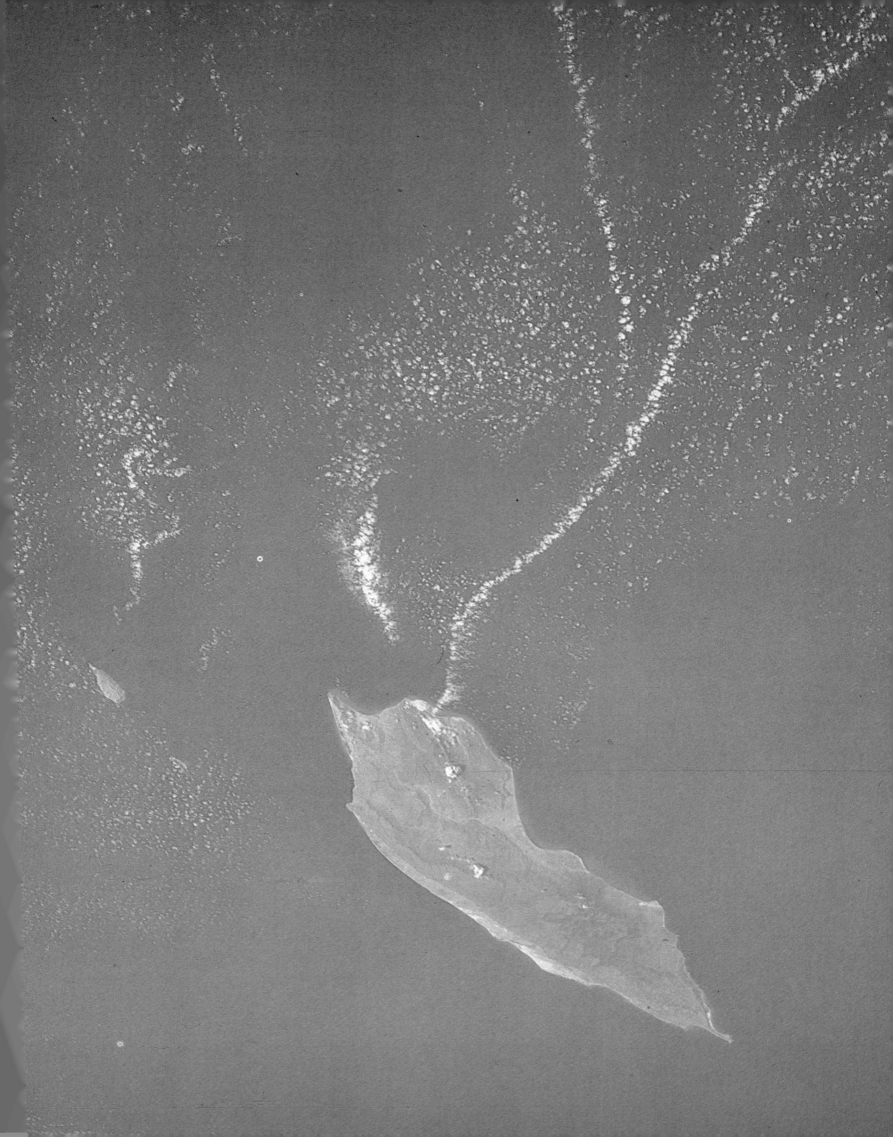

Socotra Island (Fig. 62) in the Indian Ocean off the Somali Democratic Republic is bigger and higher than Guadalupe Island, but it too stands alone in the midst of open ocean, forming an obstacle to winds. Here, rather than a von Karman street of spirals, a long thin line of clouds, known as a lee train, extends for 200 km (125 miles) downwind. It is not difficult to explain the formation of clouds behind the island. The moisture-laden southerly winds coming off the ocean are forced to rise over the sudden obstacle in their path to higher, cooler levels where the moisture condenses to form clouds. It is more difficult to account for the arrangement of the clouds in such long, well-defined trains. Von Karman vortices are not apparently developed (except rather weakly near the small island to the left), but they are probably present none the less, and may be focusing the cloud build-up along the centre of the island's wind wake. Cloud systems are never simple, and it is not always possible to explain every aspect of them. In this case, it is puzzling that the cloud train seems to be generated from the relatively low westernmost end of the island, whereas the highest point (Jebel Haggier) lies considerably further east, and appears to have generated only the tiniest puff of cloud.

Swirling eddies result wherever parts of a fluid are moving at different speeds, but to become visible they need some form of marker within the fluid. Clouds act in this way in the air; floating ice works in the same way on water. During the northern summer, the east coast of Greenland is almost free of ice, but it rapidly re-forms in autumn. Figure 63 is not a cloudscape, as it might at first seem, but a seascape. It was obtained in October 1972 when the surface of the sea was covered with a churning mass of large, old floes mixed with newly forming ice which is dispersed as a sort of haze over the surface, the individual floes being too small and thin to appear individually at this scale. Kap Broer Ruys, a rocky peninsula of the mainland of Greenland, is visible at the centre of the left-hand margin, its mountains casting the only visible shadows in the image. Close to the coast the sea is relatively static, and here large plates of ice, several kilometres across, mill around grinding against one another. Further out, towards the open sea, ocean currents make themselves felt and only newly formed ice is present, swirling around in the eddies. It is difficult to make out any general sense of movement in the image, but currents off the north Greenland coast in October generally set sluggishly towards the south. In only a few short weeks after the image was acquired the Greenland Sea would have

63 A carpet of jostling ice-rafts gyrates slowly off Greenland as the Arctic Ocean begins to freeze over (Landsat, false colour). Image scale 1 cm = 5.5 km.

frozen over to form the winter pack-ice, and the ephemeral kinetic patterns would have been locked into immobility for the long Arctic winter.

Clouds are merely the visible expression of the water vapour that is present everywhere in the atmosphere. The reason the height of the cloud base is usually sharply defined is simply because this marks the point at which the air temperature is cool enough for the water to condense. Often, particularly in high latitudes, a continuous layer or stratum is present at this level, a leaden grey ceiling which can persist for weeks, bringing depression and gloom to the wretched mortals beneath. As aircraft passengers know, such layers look much the same from above as beneath – a continuous sea of cotton wool from horizon to horizon. (The difference is that from an aircraft the clouds are usually seen in bright sunshine and look brilliant white, so that one can contemplate them benevolently as one sips a gin and tonic.)

Layer clouds, then, are an important, but not particularly interesting cloud type. Much more exciting visually are clouds which have a vertical dimension, generated by currents of rising air. To understand these fully, they need to be seen in a three-dimensional view, but images provided by standard meteorological satellites and most aerial photographs are vertical and show only the tops of cloud formations. This is an area in which Shuttle images are particularly useful because oblique low-altitude shots splendidly reveal a cloud's vertical development. Figure 64 is a case in point. Here Shuttle 2 astronauts photographed some cloud banks over the Indian Ocean, lit from behind by a low afternoon sun. The combination of sun angle and long shadows brings the clouds to life; their shapes and distribution are much more intelligible than they would be in a vertical photograph, with flat lighting.

In the foreground, the castellated, cauliflower-like tops of cumulo-nimbus clouds can be seen, caused by bubbles of warm, moist air rising high into the atmosphere. In the right foreground and background, more extensive sheets of cloud extend towards the horizon, but these too have billowy textured surfaces, also produced by convection. Locally,

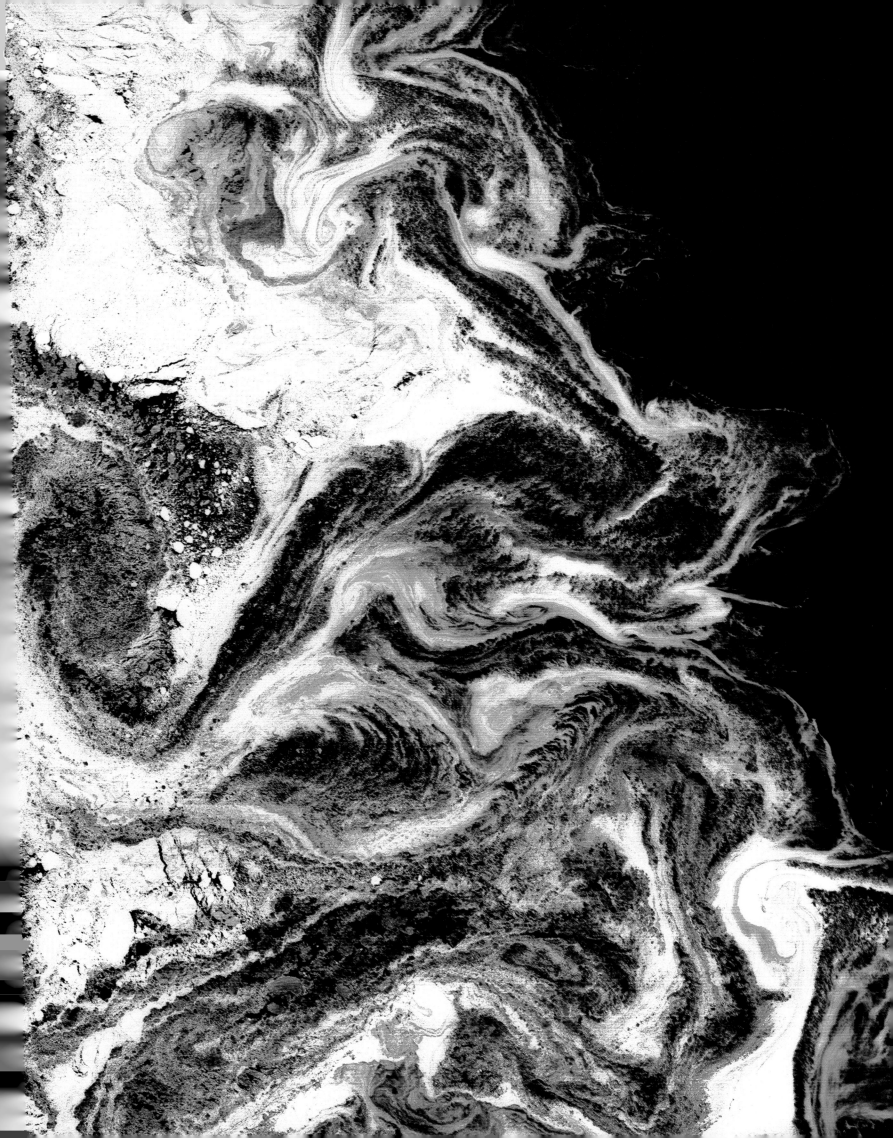

64 The low sun angle and an oblique viewpoint highlight the texture of a typical cloudscape over the Indian Ocean (Shuttle 2, natural colour). Image scale 1 cm = 15 km.

cumulus towers have punched through these layers and rear high above them. Below the clouds the air is very hazy, sufficiently hazy for the cumulus towers to cast their shadows within it.

Impressive though they are, the clouds in Figure 64 are insignificant compared with the extreme vertical development shown in Figure 65. These mushroom-shaped clouds are not the result of nuclear explosions, but the tops of powerful thunderstorm clouds bursting upwards into the atmosphere. They were photographed over Zaire, Africa, by the crew of the sixth Shuttle mission in 1983. From below, the towering cumulo-nimbus clouds that spawn thun-

derstorms are often unimpressive, since only their grey, threatening bases can be seen. Occasionally, the tops of more distant clouds can be glimpsed from the ground, and then it is possible not only to see how fearsomely high they rise, but also how quickly they punch upwards, fresh rounded crests constantly welling up.

Cumulus and cumulo-nimbus clouds develop their vertical structures as currents of warm, buoyant air (known to glider pilots as thermals) rise upwards through the atmosphere, propelled initially by the sun's warming of the land surface. At the condensation level, the cloud actually manifests itself in visible form, but the

warm air keeps rising. In the middle part of the cloud, the vertical air currents are often extremely fast, sweeping upward at 40 to 50 km (25 to 31 miles) per hour, and cause severe turbulence to aircraft flying through them. As it continues to rise, the warm air cools progressively and eventually reaches a level at which its density is the same as the surrounding air so the plume can no longer rise. At that point it diffuses out sideways to form the mushroom caps seen in the image. A double cap is visible in the largest cloud. Where strong upper atmosphere winds prevail, the top of the cloud often becomes extended downwind to form the classic 'anvil' form. Here, the symmetrical

shapes of the columns suggest small relative wind velocities and that the clouds are almost stationary. Exceptionally, clouds of this kind can reach altitudes exceeding 20,000 m (65,000 ft), but the tops of those shown in the image are probably somewhat lower.

Dramatic thunderclouds like those displayed in Figure 65 are the extreme expressions of the instability that is developed in the atmosphere when the sun shines brightly and the air is moist. A much more common, almost universal expression, are the small cumulus clouds that appear almost every day in tropical latitudes and are often seen on warm sunny days in temperate latitudes. These small white clouds with sharply defined bases and crisply rounded tops are practically synonymous with fine summer weather in the British Isles, for example, and tend to be noticeable when they pass briefly in front of the sun. Then, temperatures plummet, the breeze that was pleasantly mild moments before feels suddenly cool and goose pimples appear.

Although the sun shines uniformly over land and sea, forest and field, different surfaces do not warm up at the same rate because they have varied thermal properties. Thus, the dark earth of a ploughed field may get much hotter than a neigh-

65 Intensely convecting thunderclouds punch upward like the mushroom clouds of nuclear explosions over the forests of Zaire (Shuttle 6, natural colour). Image scale 1 cm = 4 km.

66 Gaps in the evenly-spaced puffs of cumulus cloud reveal the cool waters of rivers beneath, winding through the jungle of the Orinoco, Colombia (Shuttle 7, natural colour). Image scale 1 cm = 7.5 km

bouring one bearing green wheat, and similarly the large areas of grey roofs and roads in a town or village get hotter than the fields which surround them. From these hot surfaces, bubbles of warm air – thermals – rise and feed upwards into cumulus clouds. Experienced glider pilots can spot such clouds and the surfaces responsible for them from a considerable distance, and will use the thermals to keep their aircraft soaring for hours at a stretch.

But what happens when the surface being warmed by the sun is homogeneous, so that there are no distinct hot spots? The sea surface is perhaps the most obvious example, but the continuous green canopy

of a forest or the expanses of grassland in savannah regions are equivalent. In cases such as these, if the wind is not too strong, a remarkably regular pattern of convection is set up and the sky becomes studded with small, evenly-spaced cumulus clouds. Consider Figure 66, an area of tropical jungle in the Orinoco basin of Colombia. Here, about 30 per cent of the sky is filled with cumulus clouds which are significant for the fact they are all discrete – there is no suggestion of them merging to form a continuous sheet. The cloud base is probably at a height of about 2000 to 3000 m (6500 to 10,000 ft) and the clouds are separated by distances of the same

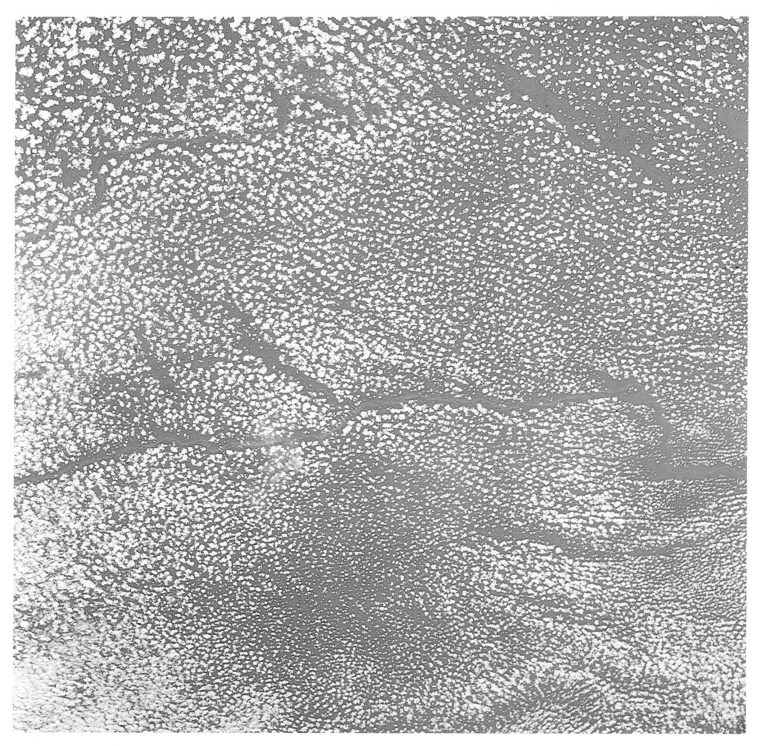

magnitude on average. There is no definite, geometrical pattern, but the clouds appear uniformly spaced, an expression of the size of the interlocking convection cells that are set up above the forest. Warm air that rises to form the cumulus clouds eventually cools and sinks down again through the cloud-free areas to rise once more in a cyclical process.

Figure 66 shows clearly the importance of rising warm air in the formation of cumulus. The only interruption in the regular distribution of the clouds is over the rivers, whose courses are marked by sharply-defined breaks. These gaps reflect the fact that there are no rising convection currents over the water because it is cooler than the surrounding forest: a traveller in a canoe would see a cloudless blue sky overhead. ('Cool' in this context is, of course, purely relative – the river traveller might actually find life unbearably hot under the tropical sun, but the air above the river is still cooler than its sur-roundings.)

On a far larger scale, the waters of the oceans are in general much cooler than the surface of dry land. Thus, coastlines can often be picked out on satellite images simply by the well-defined boundary between cloud patterns over land and sea. A striking example of this phenomenon is shown in the Shuttle image of the Liberian coast of west Africa (Fig. 67), where there are a few clouds over the cool waters of the Atlantic, but the low-lying, forested land is overhung by innumerable cumulus clouds. At the edge of the cloud belt, thin pale lines trace out the long beaches that characterize this part of the African coast.

Apart from the obvious difference between the cloud cover over sea and land, the most striking feature of this image is the way that the clouds are aligned in semi-parallel lines or 'streets'. Such cloud streets are very common; they are the result of convection taking place above warm surfaces to form cumulus cloud while a fresh breeze is blowing. In this case, a steady northeasterly trade wind was blowing over Liberia, from the land towards the sea. Liberia experiences such trade winds for periods of many months in the northern winter; the fact that they blow offshore heightens the contrast in cloud cover between land and sea, since although the clouds are carried downwind to a certain extent and may extend over the beaches, they always tend to stop abruptly over the sea. When an onshore wind blows, by contrast, cool air off the sea moves a long way inland, inhibiting convection, and thus clouds may not form over a coastal strip some 10 to 15 km (6 to 9 miles) wide, creating a well-defined band which is sunny and cloud-free, a fact which is of

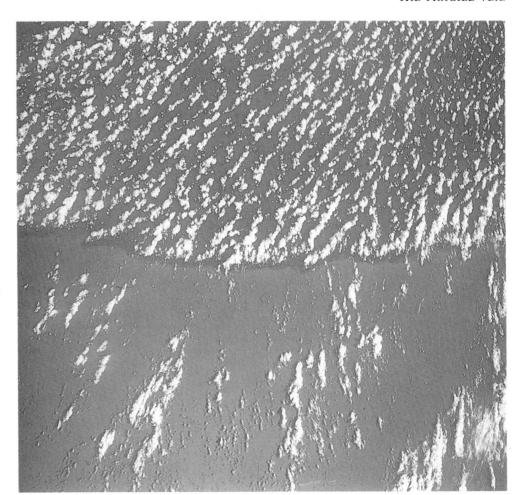

significance to some coastal resort towns.

As in Figure 66, there is a high degree of regularity in the distribution of the cloud streets in the image. It has been shown that cloud streets will form so that the distance between them is two to three times greater than the height of the cloud base above the ground. In the image, the streets are about 5 km (3 miles) apart, suggesting that the cloud base was at about 2000 m (6500 ft). The regular spacing is the result of the combination of horizontal wind currents and vertically convecting air; the cumulus clouds form in chains over the rising air, while cool air sinks in the gaps between the clouds. If one could trace the path of an individual particle in the air, it would follow a corkscrew-like track downwind, with clouds forming around it on all the vertical ascents and dissipating on the descents. As the image shows, this cork-screw motion continues for tens of kilometres and the regular spacing is governed by the diameters of the invisible, parallel corkscrews.

All the clouds illustrated so far have formed as a result of water vapour in the air condensing to liquid – a cumulus cloud is nothing more than a fog or mist of tiny suspended droplets. When first formed, the droplets are so tiny (about 0.01 millilitres) that they do not fall out of the cloud, but

67 **Crisply-defined cloud streets demarcate warm land from cool sea along the Liberian coastline, west Africa (Shuttle 5, natural colour). Image scale 1 cm = 7.5 km.**

How cloud streets are formed.

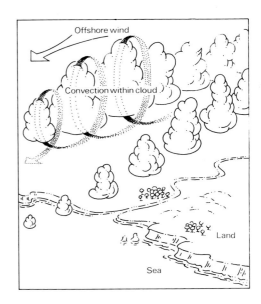

are maintained in suspension by the swirling air currents. If conditions are right, though, and the cloud evolves in the right way, enough water condenses for the minute droplets to collide and coalesce with one another, forming progressively bigger droplets. When they reach a diameter of about 0.2 mm (0.008 in), they have the size of drizzle droplets, and begin to fall out.

In thunderstorms much larger drops can form, more than 5 mm (0.2 in) in diameter, as anyone unlucky enough to be caught out in a storm can testify. In a thundercloud, large raindrops may be falling while powerful convection currents carry the smaller water droplets vertically upwards. As they rise, the droplets rapidly reach and exceed the level at which temperatures drop to freezing-point and become deeply chilled or supercooled. These droplets may not freeze until temperatures fall below $-25°C$ ($-13°F$), so that the upper parts of the cloud tend to be a mixture of water droplets and ice crystals. At the very top of large cumulo-nimbus clouds, the temperature falls to as low as $-40°C$ ($-40°F$) and only ice exists: the cloud is now 'dry'. Whenever a thunderstorm (cumulo-nimbus) cloud develops an anvil, there is a visible difference between the texture of the main body of the cloud and the anvil, reflecting the fact that the former is composed of water while the latter is made up of ice. The anvil is fibrous and diffuse, while the main cloud is bulbous and well-defined. Ice clouds tend to be much more stable and long-lasting than water clouds, and thus get shredded, sheared and extended by the upper atmosphere winds. The thunderclouds over Zaire (Fig. 65) show well-developed ice anvils, but these have not been much extended.

Figure 68 shows a quite different kind of cloud which consists of nothing but fine ice particles at high altitudes in the atmosphere, drawn out by the wind into thin, diaphanous veils. Known as cirrus clouds, these delicately drawn streaks across the sky are most commonly seen in hot or dry countries, but only because lower level clouds tend to obscure them in more temperate regions. Seen against the azure blue of a desert sky, they can make all the difference between a dreary, monotonous landscape and one that is dramatic and appealing.

Cirrus clouds appear on innumerable satellite images. The one selected to illustrate them was acquired over the Zagros Mountains in western Iran, not far from the area shown in Figure 16. Here, strong upper atmosphere winds were blowing from the northeast (the bottom left hand corner of the image), trailing a plume of fibrous cirrus over the entire

width of the area shown. In detail, the reasons for the formation of ice clouds at high altitudes are quite complicated, but in this case it may have been due to a belt of slightly moist air (bottom right) meeting a mass of drier, colder air (upper left). The little elongate shreds of cloud at right angles to the plume suggest that the two air masses are intermixing to some extent.

Beneath the clouds, the barren folds of the Zagros Mountains are crisply displayed. There are some small but distinct clumps of cumulus clouds forming over the mountains (especially at bottom left), but these are far below the level of the cirrus and nothing whatever to do with them. Viewed from beneath, the cumulus clouds would be thick enough to block out the sun, and would appear as the familiar, greyish-white looking blobs. By contrast, just as the ground can be seen through the cirrus clouds in the image, so the Sun could be seen through the hazy wisps of cloud by someone on the ground. The cirrus is so extensive that it would appear to cover much of the sky if you were immediately underneath it. Someone on the ground would also see a bright halo surrounding the Sun, a phenomenon that is totally distinctive of ice-crystal clouds. These halos, which are quite common, should not be confused with rainbows. Rainbows are always seen when looking away from the Sun, and are formed when internal reflections in myriads of raindrops bounce sunlight back towards the observer. Cirrus halos are seen when looking *towards* the Sun, and are formed by refraction of sunlight through the ice crystals of the cloud.

Apart from the churning vortices in the ice-clogged waters of the Greenland Sea, this chapter has been concerned only with clouds, since they form such an integral part of the Earth's appearance from space and cover so much of its surface. The oceans of the world also figure largely in any view of the Earth from space but, in general, they do not yield very spectacular images, since by their very nature they are flat and uniform. But this is not to say that they are unworthy of investigation. In fact, a whole new generation of satellites is being developed specifically to study the ocean surface and oceanic circulation. Because the features on the surface of the sea are relatively small-scale – most waves are less than a few metres in height – they are difficult to detect on optical images, and the new satellites are all radar-based.

68 A fragile veil of cirrus cloud trails over the barren Zagros Mountains of Iran (Shuttle 2, looking south, natural colour). Image scale 1 cm = 4 km.

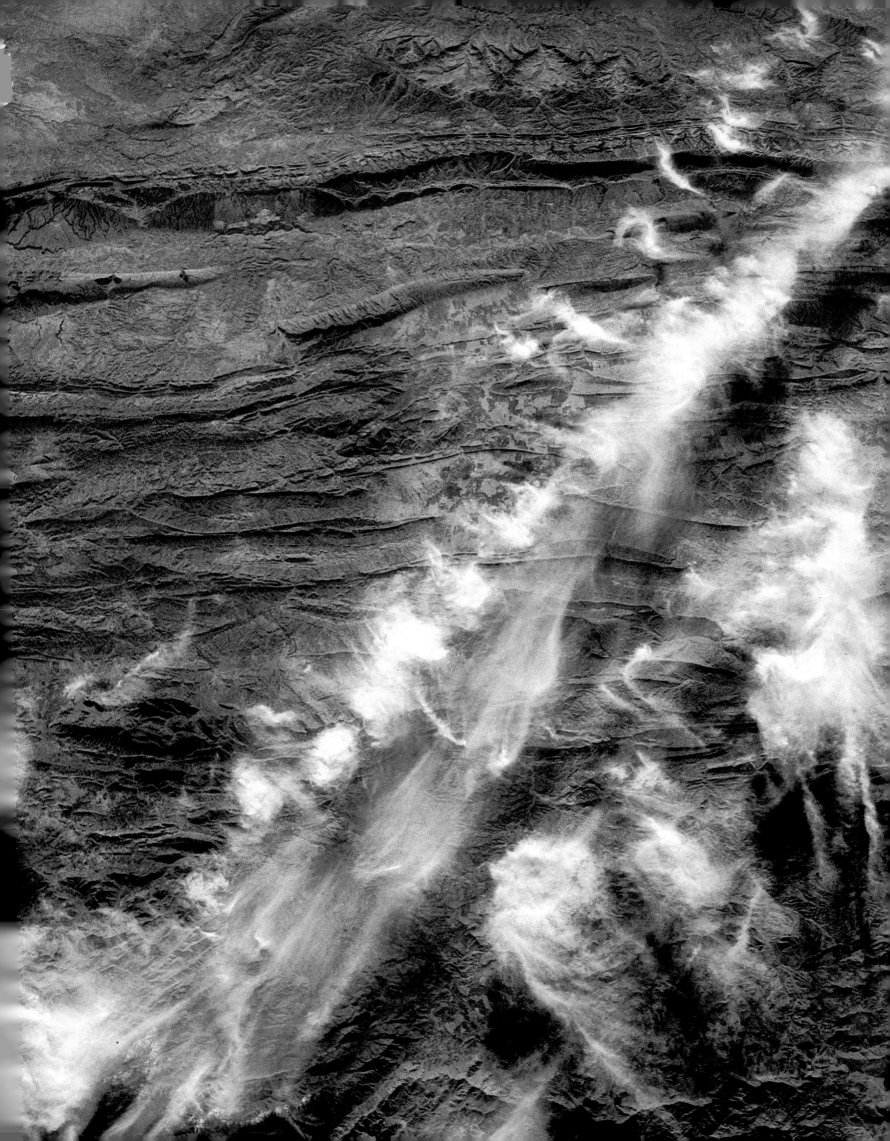

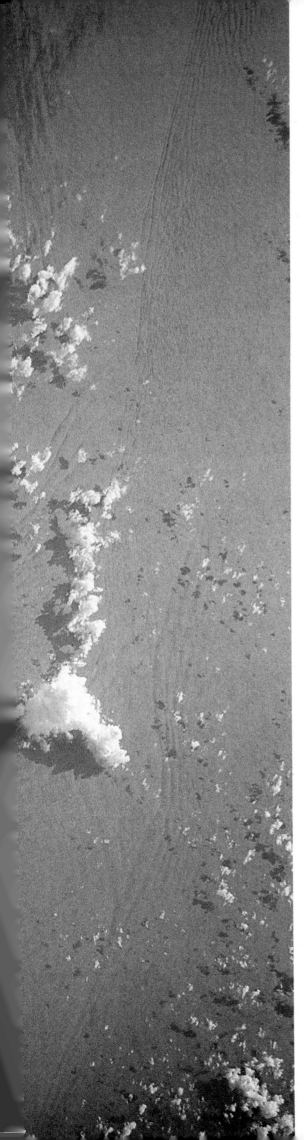

The long radar wavelengths (about 23 cm, 9 in) are excellent for detecting wave and swell patterns, and also have the advantage that precise measurements of the height of the sea surface can be taken.

Strange as it may seem, the surface of the sea is not everywhere absolutely level, since it varies slightly with atmospheric pressure and, more significantly, according to differences in local gravity. Where the local gravity is anomalously low, so the sea surface is correspondingly higher. Thus, mapping the height of mean sea-level around the world is an excellent method of mapping the Earth's gravitational field, and in this way deeply submerged mountains and ridges can be precisely delineated. Radar can also be used for more sinister purposes. A ship on the surface of the ocean leaves a V-shaped wake behind it. A deeply submerged submarine, although quite invisible itself, similarly leaves a conical wake in the water behind it which shows up clearly when the wake waves reach the surface. Thus, radar satellites have the potential for monitoring the movements of submerged submarines, if their wakes can be disentangled from all the other waves at the surface.

Radar studies apart, in some cases striking and unusual oceanographic effects can be seen in ordinary photographic images, especially when the combination of wavelength and sun angle is exactly right, or when extremely subtle details are emphasized in the reflection of the sun off water. As an example of these phenomena, a Shuttle image of part of the South China Sea off Hainan Island has been selected (Fig. 69). In the image, the land mass itself is not particularly conspicuous, being largely masked by clouds, but the white beaches along the coast appear at top left. Offshore, three separate large 'packets' of waves can be seen. It is important to emphasize that these are not ordinary waves. The packets are about 40 km (25 miles) apart, and the individual waves in each packet have wavelengths of about 1500 m (4900 ft), much larger than that of visible waves that can be seen breaking on shore. At present the origin of these wave packets is not fully understood – they may have something to do with the complex tidal movements that take place in the South China Sea and the very shallow Gulf of Tongking, but until more pictures are available covering a longer time period it will be difficult to determine exactly what they are. All over the world, similar unanticipated and puzzling wave patterns have been detected on Shuttle imagery, which seems sure to provide oceanographers with a rich and rewarding new field for research.

69 An oceanographic puzzle: packets of large waves off the coast of Hainan Island, China (Shuttle 7, natural colour). Image scale 1 cm = 5 km.

7 A Pattern of Islands

The idea of an island, a piece of land entirely surrounded by sea, is very appealing. This may to some extent be because it is so much easier to identify with a wave-girt scrap of land than with other more amorphous entities defined by political or cultural boundaries. Thus, tiny Pitcairn Island conveys a real sense of identity, whereas one can only think of much larger Paraguay, for example, as a rather vague part of an enormous continent.

Our benevolent feelings towards islands are probably reinforced by the fact that most of us tend to associate them with warm tropical seas, white beaches and sparkling lagoons: the ideal places for holidays. This image is bolstered by a vast literature which emphasizes the romance of tropical islands, and by the travel industry which works tirelessly to convey us as cheaply and as uncomfortably as possible to the nearest island that can be made to measure up to the required image. While there are a remarkably large number of islands in tropical and subtropical regions that do match the exotic concept,

there are many more at higher latitudes that are in fact quite the opposite: bleak, frigid and desolate. Our minds turn much more often to the charms of Tahiti or Fiji than they do to the South Shetlands or to Bouvet Island in the far South Atlantic. From the perspective of an orbiting Shuttle, photographing islands seems particularly attractive, precisely because the view-finder encapsulates them completely.

Although islands come in an infinite range of shapes and sizes, from continents to ephemeral sand-bars, we usually think of them as being fairly small-scale. There are three basic physical types: young volcanic islands, that may poke abruptly up from deep ocean and reach heights of thousands of metres; atolls, which generally do not rise more than a few metres above sea-level; and 'continental' islands, areas of old rocks of continental origin that are usually part of a continental shelf that has been drowned by a rise in sea-level. There are links between these three groups, as we shall see.

Tenerife (Fig. 70) is a high volcanic island, reaching 3718 m (12,198 ft), so it lies firmly in the first category, but it is also one of the Canary Islands, some of which come very close to the northwest coast of Africa. Some geologists believe that at least the inner islands of this group are built on rocks forming part of the African continent. Politically, Tenerife is part of Spain

70 Tenerife, Canary Islands, one of Europe's most popular tourist destinations. Teide volcano, highest mountain in Spain, dominates the island (Shuttle 9, natural colour). Image scale 1 cm = 5 km; map scale 1 cm = 20 km.

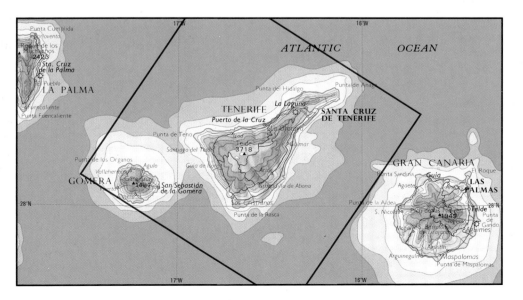

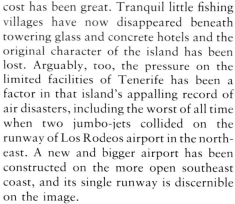

and surprisingly contains Spain's highest point – Pico de Teide, the snow-capped volcano that dominates the island. Although Teide is classified as an active volcano, and there have been eruptions within this century, this has not inhibited the massive development of tourism on the island during the last two decades. Indeed, the spectacular volcanic scenery of Teide and the Las Cañadas caldera that encircles it on the south provide the high spot, literally and metaphorically, for most vacationers visiting Tenerife.

In many ways, Tenerife and the other islands of the Canary group are ideally suited to tourism: the climate is mild in winter and hot and sunny in summer, the scenery is superb, and, above all, the islands are only a short jet flight away from the crowded cities of northern Europe. British, German and Swedish tourists in particular throng to Tenerife in spring and summer to renew their acquaintance with sunshine and easy living after the long, grey northern winter. The climate is so good and the air so clear that an important astronomical observatory has been constructed on a crest of the mountain. The island has disadvantages, though. Since Teide volcano is active, Tenerife is actually 'growing' in geological terms, which means that the island is ringed with steep lava cliffs against which the heavy Atlantic surf pounds relentlessly. (The white surf-line is well seen on the northwest coast, facing the prevailing winds.) As a result, good beaches are few and far between and some artificial ones have had to be constructed. Moreover, although tourism has undoubtedly brought wealth to what was previously a rather poor island, the

71 Sunlight gleams off the waters of the Pacific Ocean surrounding Molokai, Maui, Lanai and Kahoolawe, members of the Hawaiian group (Shuttle 7, natural colour). Image scale 1 cm = 6 km; map scale 1 cm = 50 km.

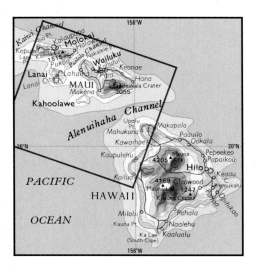

cost has been great. Tranquil little fishing villages have now disappeared beneath towering glass and concrete hotels and the original character of the island has been lost. Arguably, too, the pressure on the limited facilities of Tenerife has been a factor in that island's appalling record of air disasters, including the worst of all time when two jumbo-jets collided on the runway of Los Rodeos airport in the northeast. A new and bigger airport has been constructed on the more open southeast coast, and its single runway is discernible on the image.

Another volcanic island paradise is the Hawaiian group, in the middle of the Pacific and about as far from a continent as it is possible to get. The main island is Hawaii, the Big Island featured in Figure 6. Figure 71 shows some of the islands to the west, Molokai, Maui, Lanai and Kahoolawe. Trade winds laden with moisture cause clouds to build up over the northeastern flanks of the islands. Beneath the clouds, the luxuriant forested valleys provide the ideal tropical island atmosphere, with waterfalls, brilliant flowers hanging from rampant vegetation, and brightly-coloured birds calling from the treetops and flitting overhead.

Since these islands are so much further from large mainland cities than the Canaries, they have not been subjected to quite such intensive development, but changes have been profound nonetheless. Traditionally, the islanders grew sugarcane and pineapples and this continues to some extent. In particular, Lanai's 60 sq km (23 sq miles) are almost entirely given over to pineapple growing, and the whole island is in effect under the dominion of the American Dole fruit company. On Maui, however, large tracts of what was agricultural land have been sold for development, with wealthy Americans buying homes and vacation condominiums. Once established on the island, with their new houses offering wonderful views of the volcanic slopes and the azure sea beyond, the newcomers fiercely resist further developments that might encroach on their land and spoil their views with unsightly resort hotels. Kahoolawe is undoubtedly the most natural and unspoilt of the islands, but only because it is a bombing range for the US navy and is littered with live explosives.

Although none of these islands is volcanically active at present, they are all nothing more than the eroded summits of volcanoes similar to the still-active Mauna Loa on Hawaii, sticking up above the surface of the sea. On Maui, the 3055-m (10,023-ft) volcano Haleakala shows an abundance of features suggestive of recent activity – crisply-shaped cinder cones and fresh lava flows. The volcanic characteristics of the Hawaiian group provide an important clue to understanding the evolution not only of these islands, but also of the myriads of coral atolls that dot the Pacific and Indian oceans.

Traced westwards, the islands in the chain become geologically older and their mountains lower until the end of the chain is reached in distant Midway atoll (Fig. 21), which is completely flat and has no visible rock on its surface. This island appears to be made of nothing but coral, but drilling has shown that lavas some 17 million years old are present deep underneath the coral. The Hawaiian chain, then, consists of a line of volcanoes that gets progressively younger and higher from west to east, culminating in the active giants of the Big Island. Geologists attribute this regular progression to the movement of the Pacific Ocean crust over a stationary 'hot spot' deep in the bowels of the Earth, which has in effect burned its way through to the surface as ocean crust moved over it, causing volcanic activity.

Charles Darwin would have been delighted by this explanation. When he visited the Pacific, during the voyage of the *Beagle* (1831–6), he was puzzled by the vast numbers of low-lying coral atolls he saw. He knew well enough that reef-building corals can only live in shallow water that is less than 20 m (65 ft) deep. How then, he wondered, could there be so many coral atolls throughout the deep waters of the south and west Pacific? The Shuttle image of the Society Islands, which Darwin would have given his right arm for, provides a magnificent illustration of his solution to the problem. Figure 72 encapsulates the story of the Hawaiian island chain on a smaller scale. At bottom right are the islands of Tahaa and Raiatéa; they are old, deeply eroded volcanoes, fringed by a single coral reef. Northwest of them lies Bora-Bora, a much smaller and more ragged remnant of a volcanic island. North of Bora-Bora lies Tupai, merely an atoll now, but undoubtedly originally a volcanic island. Darwin perceived clearly that there was an intimate relationship between volcanic islands, the reefs that fringed them, and coral atolls. The myriads of coral atolls formed originally as fringing reefs around volcanic islands and the islands were subsequently gradually eroded away, or subsided beneath the sea. Rocks above sea-level are eroded extremely fast, particularly in tropical climates, but below sea-level the rate of erosion is much, much slower. Thus, once an island has been planed down to sea-level, or just below the level at which wave action is effective, it forms a sort of permanent platform on which coral reefs can thrive. If, for

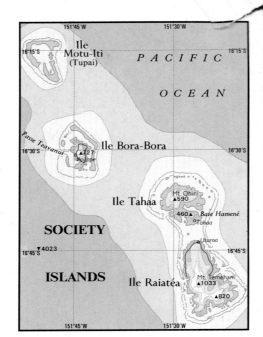

72 Tahaa, Raiatéa, Bora-Bora and Tupai in the French Society Islands. The islands remain strikingly beautiful, although the carefree culture of their original inhabitants has perished (Shuttle 8, natural colour). Image scale 1 cm = 3 km; map scale 1 cm = 10 km.

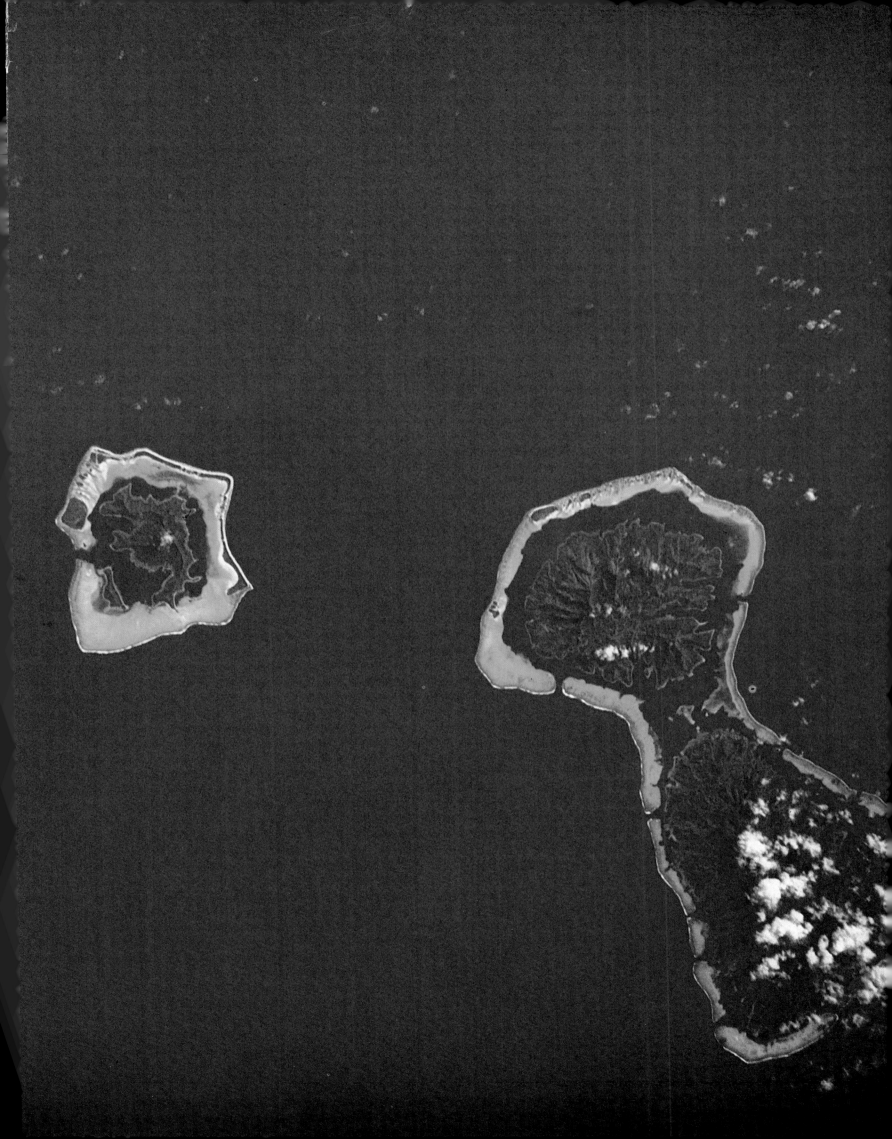

73 Canton atoll, Pacific Ocean. The vivid lagoon and encircling reef are spectacular from space but the atoll is hot, dry and deserted and living on the island would be grim (Shuttle 9, natural colour). Image scale 1 cm = 3 km; map scale 1 cm = 2.5 km.

geological reasons, the volcanic island should sink of its own accord, the reef could generally keep on building fast enough to keep itself just below the surf-line, at the ideal level for coral growth.

The shallow water of the lagoon within Bora-Bora's fringing reef has a wonderful turquoise colour which indicates shallow, crystal-clear water, a paradise for swimmers and skin-divers. From the surface the prospect is equally attractive, but the tiny 'X' of airport runways on the northwestern corner of the fringing reef reveals that this atoll is far from being a tranquil, tropical Eden. Lying downwind from the famous island of Tahiti, in the French Society Archipelago, these islands form part of a sub-group known as the Îles sous le Vent, or Leeward Islands. All islands in the group are now heavily commercialized, with facilities ranging from luxury hotels to holiday camps of the Club Méditerranée type.

Although Tahiti is today a colourful mixture of French and Polynesian cultures, it used to have its own unique social structure which was for a time the envy of the more 'sophisticated' Western World. During the eighteenth century, philosophers were becoming aware of, and depressed by, the problems that civilization brought in its wake, and a school of thought grew up which argued that man had only been corrupted and debased by the advances in civilization that he was so proud of. Contemplating these problems, the philosopher Jean-Jacques Rousseau (1712–78) conceived the idea of the 'noble savage', a species of man who had grown up away from the corrupting influences of civilization and, furnished by nature with all he required, had developed a truly noble culture, unsullied by greed, envy or hate.

When Tahiti was discovered in 1767, reports began to filter back to Europe of a spectacularly beautiful group of islands, surrounded by sparkling seas teeming with easily caught fish, and with a cornucopian abundance of fruit and flowers. It was seriously thought that this might indeed be some sort of terrestrial Garden of Eden. When it was further reported that the islands were peopled by happy, healthy individuals, who had little need to work and who lived open, innocent, communal lives, free from the social ills that plagued Europe, it seemed that the noble savage had indeed been found. This pleasant illusion did not survive long, however. Not only were the islanders far from being truly noble, since they happily engaged in violent intertribal warfare and bloodshed, but also whatever they had that was good and desirable rapidly crumbled after exposure to the outside world. The impact of western culture on the islands was fatal,

literally and metaphorically.

Captain Cook, the great English navigator, visited the islands several times between 1769 and 1779. His seamen, and those of the other vessels that came to the island, revelled in the charm and grace of the Polynesian women, and gave them venereal disease. Other diseases spread like wildfire through the islands in later years, almost annihilating the indigenous population. These physical miseries were later capped by introductions that further undermined native culture – including missionaries who arrived from Europe to encourage the islanders to abandon their former hedonistic ways, wear European clothes, and give up their 'savage' customs. Although the traditional way of life has long gone from the Society Islands, physically they remain as spectacularly beautiful as ever. If only the Shuttle could photograph some similar island, somewhere, that had not yet made its way onto airline schedules.

Although one mourns what has happened to Tahiti, there are still many islands in the Pacific that have not been transformed by man. The trouble is that such islands generally have little to offer other than postcard views of white sand, palm trees and sparkling water. Thousands upon thousand of atolls fit this description, but while many would provide a delightful setting for a week's swimming or fishing, they have little else to recommend them. Our mental concept of coral islands tends to be coloured by descriptions of steep, lush mountains like those of Tahiti. Most atolls, by contrast, are flat, rocky scraps of land with little fresh water and rather restricted vegetation, surrounded by formidable reefs against which the ocean surf breaks constantly and dangerously. Canton atoll, 3000 km (1875 miles) northwest of the Society Islands, is typical of those that are rather less than idyllic (Fig. 73). Superficially the atoll looks most attractive, a ring of white sandy beaches enclosing a translucent lagoon. In fact, it is so inhospitable that it has no permanent habitation, though the large airfield in the northwest corner shows that it has not been entirely unmarked by man.

Canton was discovered early in the nineteenth century, its name being taken from that of an American whaling ship wrecked on the island in 1854. Over the next few decades American companies extracted guano from the island, but in 1889 Britain claimed it for a trans-Pacific cable station, and then, in the 1930s when air routes over the Pacific were being pioneered, it became important as a fuelling stop. From 1939, the island was administered jointly by Britain and the USA, with the exception of 1942–3 when

the Japanese were in occupation.

In 1979, Canton became part of the independent republic of Kiribati, a strange little country comprising 33 atolls scattered over 5 million sq km (2 million sq miles) of the Pacific; the Gilbert group in the west and the Phoenix group in the east. Collectively these islands have a land area of 684 sq km (264 sq miles). Canton forms part of the Phoenix group and it is significant that the entire group is almost uninhabited, clear evidence of how barren the islands are. On Canton, the largest of the group, there are only 9 sq km ($3\frac{1}{2}$ sq miles) of land above sea-level, and such soil as there is is too poor to support useful crops. Being only $2\frac{1}{2}°$ from the Equator, the sun beats fiercely down, although cool winds usually keep temperatures depressed to below 30°C (86°F). Since the island is completely flat, there are no hills to attract clouds and rain and there are often long periods of drought. In short, once the guano deposits had been exhausted in the nineteenth century, Canton has only been of use as a way-station between places with more to offer.

Half a world away from Canton, but less than an hour's orbiting time, the Shuttle 9 photograph of Farquhar atoll in the Indian Ocean (Fig. 74) has some striking similarities – the same dazzling white beaches, the same turquoise lagoon surrounded by the same intensely azure blue sea. While it also shares many of the same problems as Canton, Farquhar is not quite so barren. The island was discovered in the sixteenth century by the Galician navigator Juan de Nova, after whom it was first named, but it was eventually incorporated into the British colony of Mauritius and it takes its present name from the first governor of the colony. Subsequently, the island was administered from the Seychelles, and after a period as part of the British Indian Ocean Territory, it became part of the independent republic of the Seychelles in 1976.

Like Canton, the atoll rises only a metre or two above sea-level, but there is sufficient soil to support substantial coconut plantations and a working population of between 60 and 100 spend periods of a year or 18 months on the island to tend and harvest the crop. The chief product is copra, the dried white meat of the nut from which valuable coconut oil is obtained. The small dirt airstrip used for ferrying personnel and supplies to the island can just be seen parallel to the beach at top left. Fish are also abundant in the lagoon and the ocean. Although more richly endowed by nature than Canton, Farquhar has its own disadvantages that Canton is spared. It is located further south than Canton (10°) and lies within hurricane latitudes, fiercesome storms periodi-

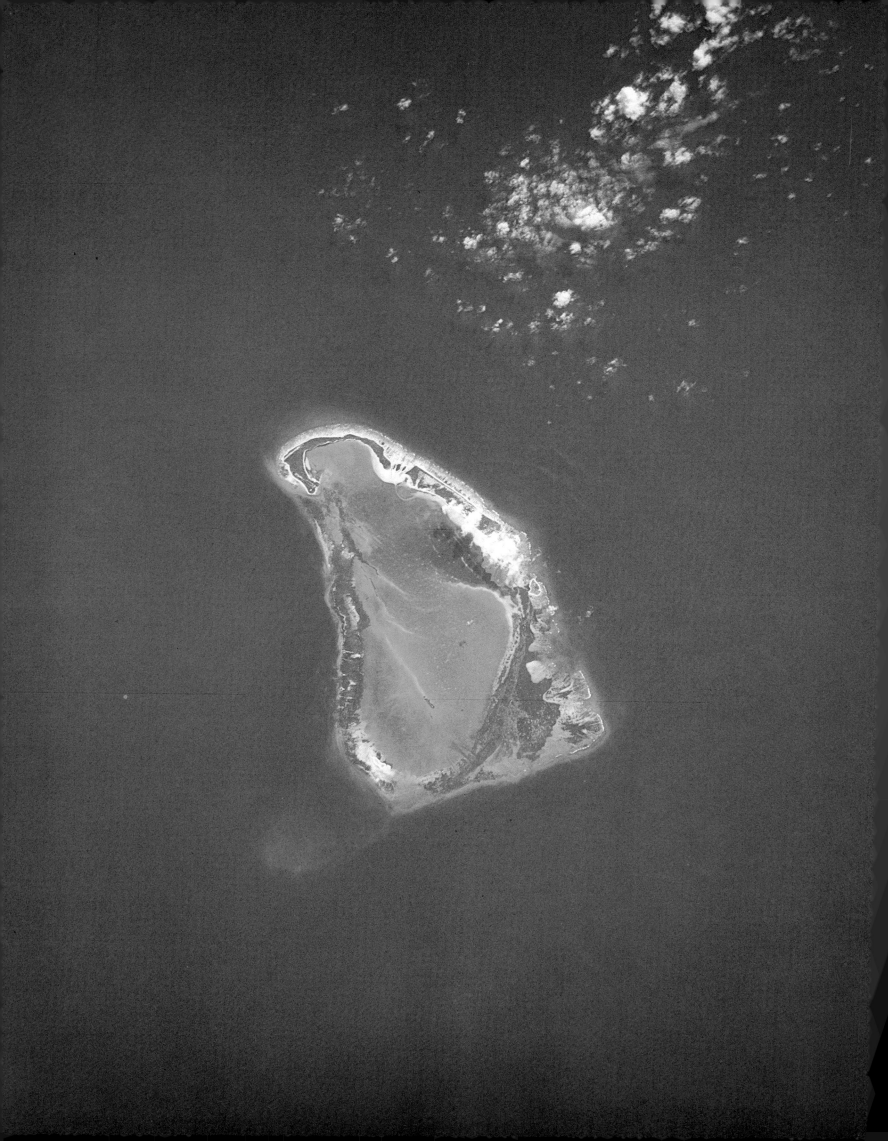

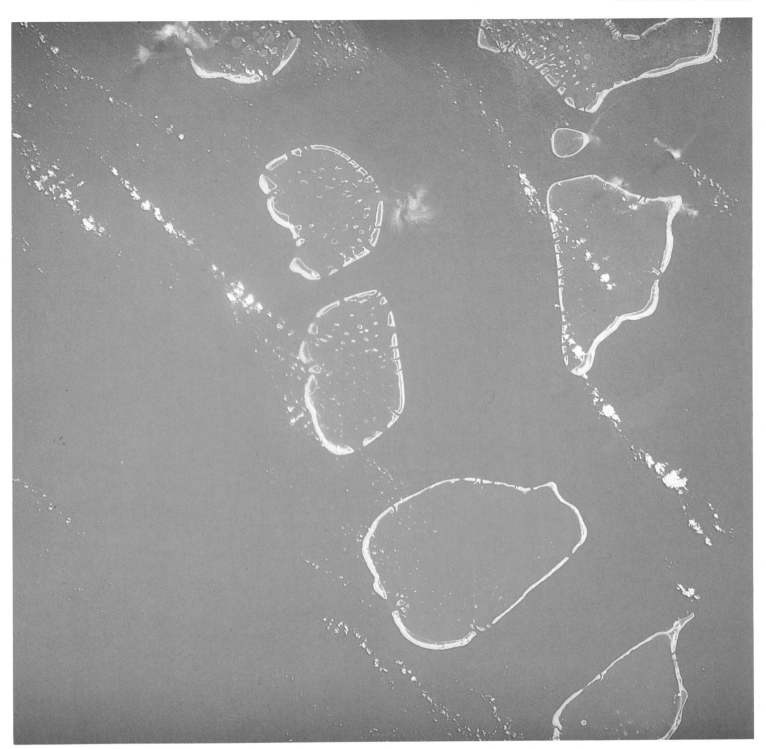

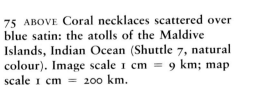

75 ABOVE Coral necklaces scattered over blue satin: the atolls of the Maldive Islands, Indian Ocean (Shuttle 7, natural colour). Image scale 1 cm = 9 km; map scale 1 cm = 200 km.

74 LEFT Coconut plantations provide a means of subsistence for about a hundred people on exquisite Farquhar atoll in the Indian Ocean (Shuttle 9, natural colour). Image scale 1 cm = 2 km.

cally sweeping across the island and devastating the coconut plantations. Damage is particularly extensive because the island is so low-lying that it has no defence against the ocean surge raised by the hurricanes. At least four vessels have come to grief during storms, claimed by the fringing reefs that look so harmless and attractive on the image.

Although its coral reefs are so similar to Canton and thousands of other atolls, Farquhar and the other islands of the Seychelles are highly unusual because they are almost the only coral islands that are not built on old volcanoes. On the Seychelles themselves, granites typical of continental crust are exposed in the hills and they probably underlie Farquhar as well, though no drilling has been done. It seems likely that these unusual islands are based on small fragments of the African continent that split off when the processes of continental drift ruptured Madagascar from mainland Africa.

Coral atolls are amongst the features most commonly photographed by astronauts orbiting the Earth. This is probably partly because they have a vested interest in them, since they are always thinking of possible alternative landing sites for the Shuttle. In the future, launches of the Space Shuttle will be made from the Vandenburg Air Force Base in California, westwards over the Pacific. Hao Island in the Tuamotu Archipelago (east of the Society Islands) is being investigated as a possible site for an emergency landing in case of an engine failure when the Shuttle is being put into orbit. Apart from such considerations, though, atolls also shine out against the dark ocean background like brilliant jewels, and are natural subjects for photography. No group of islands is more beautiful from space than the Maldive archipelago, which appears as a series of pearl necklaces against a background of blue satin (Fig. 75).

Like Canton and Farquhar, the Maldives are hoops of coral reefs built up on submerged rocky basements, but the Maldives are different in some respects. The atolls are very much larger – Kolumadulu atoll (centre bottom) is 50 km (31 miles) across, though its reef is still only a slender thread. Their lagoons have the same blue tone as the Indian Ocean, suggesting that they are much deeper than the brilliant turquoise lagoons of Canton and Farquhar. On the other hand, they must be much shallower than the surrounding ocean since coral heads dot the lagoons of the northern atolls (Mulaku and Nilandu), seeming to mimic the strings of puffy clouds that drift overhead. Large atolls like the Maldives present some problems of interpretation. Why should

such large, closed structures exist, with not a single scrap of solid rock anywhere? It is easy enough to imagine a small atoll forming around a volcanic island which eventually erodes away, but nothing breaks the surface in the Maldives except the exquisite loops of coral.

Underlying the Maldive archipelago and stretching for some thousands of kilometres southward from the west coast of India is a great submarine mountain range known as the Chagos-Laccadive Ridge, which comes near to the ocean surface but never quite breaks it. During the Ice Age, however, when vast amounts of water were locked up in the continental ice-sheets that held much of the world in their frigid grip, scientists believe that sea-level stood no less than 130 m (430 ft) lower than it does at present. Thus, for hundreds of thousands of years, parts of the Chagos-Laccadive Ridge stood at or close to sea-level, and it is thought that the coral reefs first became established at that time. As the water depth increased with the melting of the ice-caps, so the coral built upwards and outwards, keeping pace exactly with changes in sea-level. Of course such global changes in sea-level must have also affected every other island in the world, so the history of reef and atoll building is by no means simple. The Maldive atolls may be so large simply because they have been in building much longer and have a broader, flatter platform to build on than the smaller atolls.

Such considerations are of little practical importance to the inhabitants of the Maldives, of whom there are only some 140,000 thinly distributed over about 210 inhabited islands. Like the people of Farquhar, they are largely dependent on coconuts and fish, though they also raise breadfruits, mangoes, pawpaws, plantains and pumpkins.

While they are unusually large, the Maldives are by no means unique. From space, they are almost indistinguishable from the atolls of the Tuamotu group, for example. But such atolls are confined to the Indian and Pacific oceans; there are none in the Atlantic. It is not obvious why this should be so, since the Atlantic is connected to both the other great oceans, but it seems that the enormously successful reef-building coral organisms of the Indian and Pacific oceans are unable to stand the cold temperatures of the waters around the Cape of Good Hope and Cape Horn, and have thus been unable to migrate into the Atlantic. There are corals in the Caribbean, but the reefs they build do not approach those of the Indian and Pacific oceans in size, or in the variety of their coral species. Had the Panama Canal been excavated at sea-level, this would surely

have resulted in some profound changes in the Caribbean.

It is difficult to select an island that is typical of the Caribbean, but Figure 76 showing one of the Bahamas is both attractive and informative. There is almost as much to see under water as on land in this photograph, which was taken on the very first Shuttle mission in 1981. The thin, fish-hook of an island is Eleuthera, some 128 km (80 miles) long but only 9.5 km (6 miles) wide at its broadest. Eleuthera was one of the first islands in the Bahamas to be settled when the British came over from Bermuda in 1647. The permanent population is quite small, numbering only a few thousands, but many tourists come to the island to enjoy the clear blue waters and white beaches that the image displays well. An airstrip at Deep Creek is visible on the widest part of the island.

Eleuthera Island itself is little more than a sand-bar rising a few metres above sea-level and separating the deep, dark-blue water of the Atlantic from the lighter-coloured shallows of the Great Bahama Bank to the west. (The name 'Bahama' comes from the Spanish *bajamar*, meaning shallow sea.) Exuma Sound, a tongue of deep water from the Atlantic on the west of the island, meets the shallows in a well-defined curve. The water over much of the Bank is not more than 10 m (33 ft) deep, so details of the topography of the bottom can be seen through the crystal-clear sea. Parts of the Bank are extremely shallow, only a few metres deep, and here the white sand on the sea-floor shows plainly. An intricate pattern of channels and bars has been cut in the sands by powerful tidal currents sweeping back and forth from Exuma Sound over the shallow bank.

In this image, the Space Shuttle has provided an unusual view of a natural chemical laboratory, one of the few places on Earth where limestones can be observed in the process of formation. The shallow water is very warm, and becomes extremely salty. Crystals of a calcium carbonate mineral known as aragonite precipitate and form into small spherical grains, oolites, as the tidal currents swirl back and forth. The white sand of the Bahamas in fact consists almost entirely of carbonate material; the islands are so far from the mainland that there are none of the quartz grains that are usually found in beach sands, and there is no clayey mud either, so the sands remain beautifully white. Lithification of the carbonate sands forming here produces a kind of rock known as an oolitic limestone, similar to the well-known Portland limestone of Britain.

The processes at work in Figure 76 are a total contrast to those displayed in images

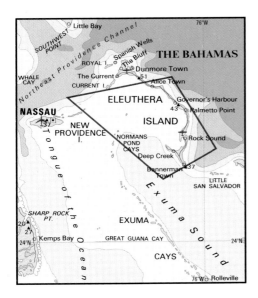

76 A submarine limestone factory: white carbonate sandbanks off Eleuthera Island, Bahamas (Shuttle 1, natural colour). Image scale 1 cm = 3 km; map scale 1 cm = 35 km.

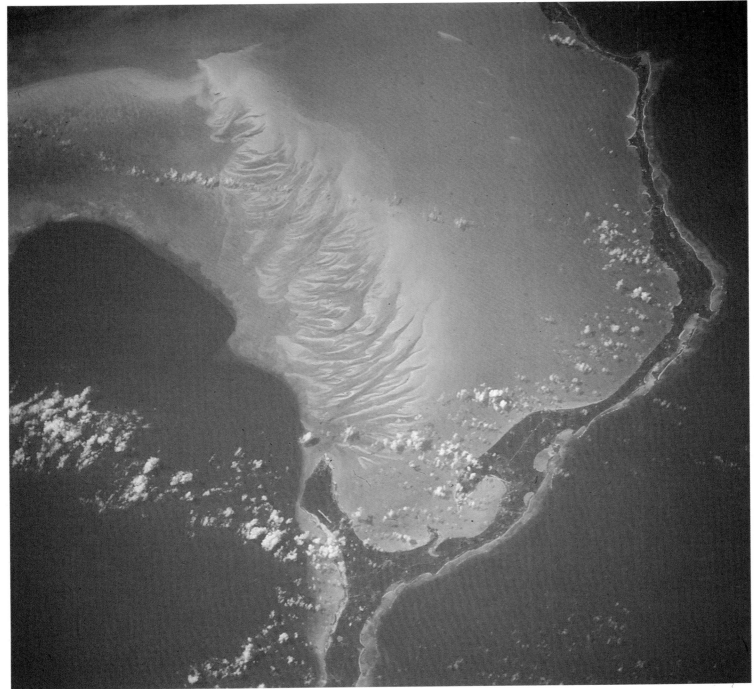

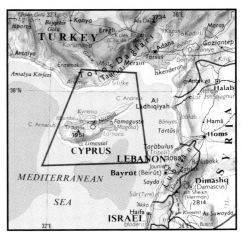

77 Apparently serene seen from space, the strife torn island of Cyprus is encapsulated in this Shuttle image which emphasizes its proximity to Turkey (Shuttle 2, looking south, natural colour). Image scale 1 cm = 7 km; map scale 1 cm = 100 km.

of atolls. Though both produce limestone, coral limestone is formed organically. The carbonate sands of the Bahama Bank, on the other hand, are formed by largely inorganic processes. In fact, the water over the Bank is so warm and salty that it is inimicable to most kinds of organic life, except for some hardy shellfish that burrow into the sand. There are some corals around Eleuthera, but they are few and found only on the *eastern* coasts, where the water is cool and not over-salty.

Islands made up of portions of land which are fundamentally continental in character tend to be much larger and more diverse than the volcanic or coral islands illustrated so far. Such islands have usually resulted from the submergence of part of a continental area and are formed of the higher land that was not drowned by the sea. Britain and Ireland originated in this way, as did the numerous islands in the Baltic (Fig. 10). More rarely, chunks of larger masses are detached by the processes of continental drift, and the gap between the two is filled by new ocean crust. Madagascar became detached from Africa in just this way.

Cyprus (Fig. 77) is an example of an island formed by submergence. There is no mistaking the distinctive outline of the island in the image, set against the foreground of the southern Turkish coast. Its shape has been largely determined by a pair of ridges, part of two much greater mountain arcs which sweep from the Asian mainland to Crete. Close to the north coast the Kyrenia mountains run in an east-west line, highlighted by a few small clouds forming along the backbone of the range. To the south lies the Mesaoria Plain, and

N

beyond this rears the major Troodos mountain massif, crowned by the 1951-m (6400-ft) Mount Olympus. Both the Kyrenia and Troodos ranges are forested and show up in dark tones, while the lower ground is lighter coloured and more extensively cultivated.

Cyprus is of outstanding interest to geologists, since the Troodos massif contains a rare assemblage of rocks typical of ocean crust which have here been thrust up over continental rocks. Unfortunately, however, its tragic cultural and political history commands much more immediate attention. Located in a strategically vital corner of the eastern Mediterranean, 60 km (37 miles) south of Turkey, 100 km (62 miles) west of Syria and 400 km (250 miles) north of Egypt and a similar distance from the Greek islands, Cyprus has been occupied or colonized by numerous powers in its long history. In the earliest days, the Phoenicians and Romans sought copper on the island, which gets its name from that of the metal (*cuprum*), but for most of its history Cyprus has been inhabited by people of Greek descent. In 1571, however, Ottoman Turks invaded the island, and continued in occupation until 1878 when the British took over. It was the initial Turkish occupation that sowed the seeds of the bloodshed that continues to this day. This was exacerbated by a fresh Turkish invasion in 1974, which resulted in the division of Cyprus. Today, Greek Cypriots make up 77 per cent of the population; the remainder are of Turkish descent. This situation is slightly paradoxical, since Cyprus is much nearer the Turkish mainland than the Greek.

Cyprus was formally annexed to Britain in 1914 as part of the complex supportive alliances formed at the outbreak of World War I. After years of trying to suppress the Greek-Turkish hostility, and in particular the Greek Cypriot demand for *énosis*, or union with Greece, the British granted independence to the island in 1960, but under the Treaty of Establishment of the Republic of Cyprus Britain retained 99 sq miles (256 sq km) of military enclaves on the island. These are bases for some 10,000 servicemen supporting NATO in southeastern Europe. At present, although active hostilities have been subdued with the help of the UN, the island remains unhappy and divided. Greek Cypriots have been denied their dreams of *énosis*, and the northeastern part of the island is under Turkish administration. Tourism, which once brought wealth to the island, has been stifled and there seems little immediate prospect of a brighter future.

Over the years, innumerable images of islands have been acquired from one

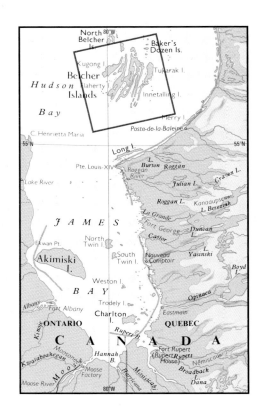

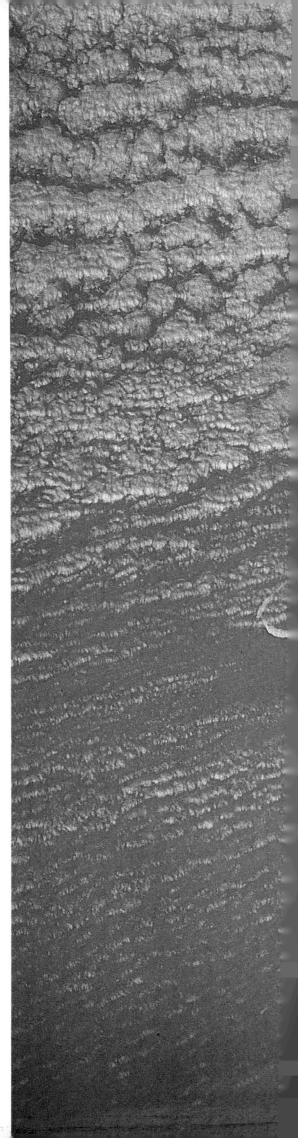

78 Appearing more like silky white ribbons than solid rock, the Belcher Islands in Hudson's Bay, Canada, are seen through a thin veil of cloud (Shuttle 9, natural colour). Image scale 1 cm = 5.5 km; map scale 1 cm = 70 km.

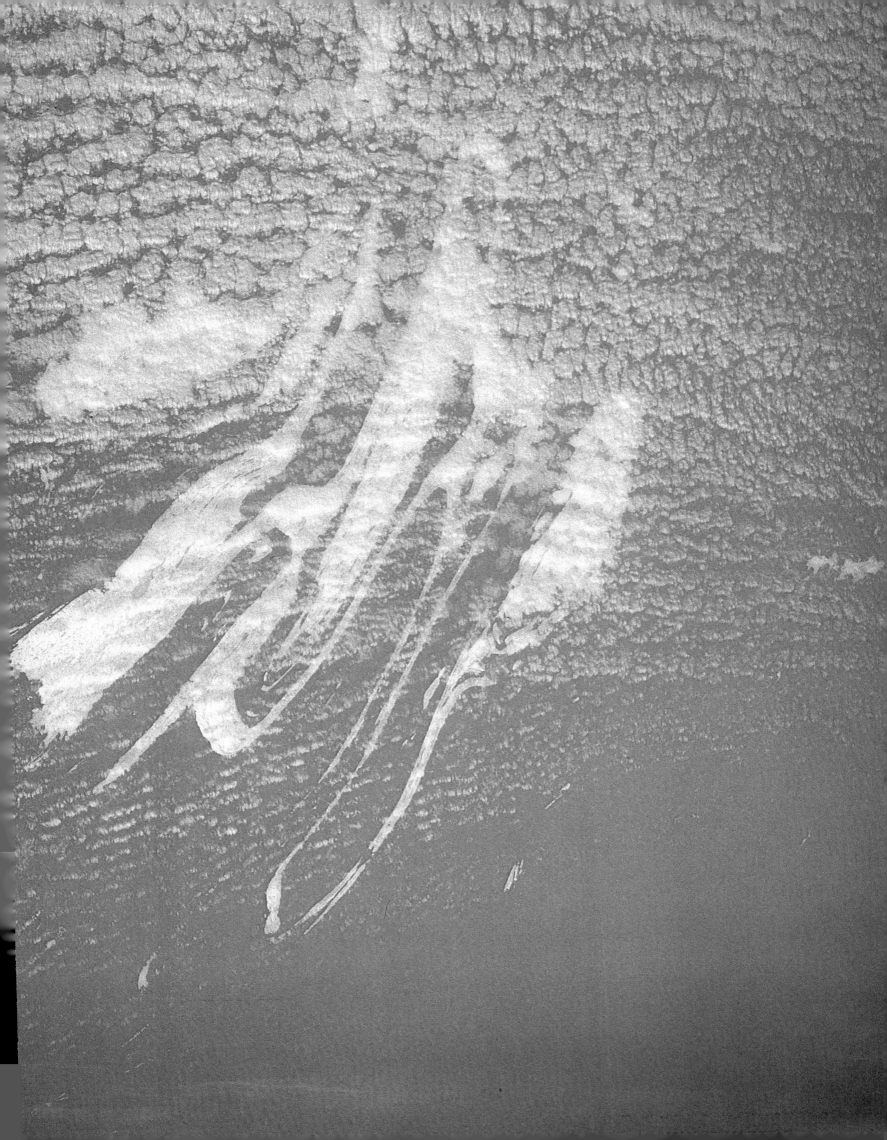

spacecraft or another. Of these, none is more striking and unusual than Figure 78, which has much of the quality of a trompe l'oeil painting. At first sight, the swirling white ribbons look like the sweeping strokes of an artist's paint-brush over a blue canvas, but the gauzy veil of clouds through which the outlines show demonstrate that this is indeed a photograph from space. The crisply defined loops and whorls look so insubstantial that they could well be something floating on water which has been churned up by local currents – ice perhaps, or froth. Surprisingly, the image is actually a Shuttle 9 view of the rock-solid Belcher Islands in Hudson's Bay, taken in December 1983 when the islands were snow-covered.

These unusual, low-lying islands extend over about 13,000 sq km (5000 sq miles), and have a land area of about 2800 sq km (1080 sq miles). Their ribbony shapes are the result of the submergence of an eroded sequence of thinly-bedded, folded sedimentary rocks, of which the harder, more resistant emerge above sea-level. Perhaps the best way to grasp their formation is to imagine a series of rocks like those of the Zagros Mountains (Fig. 46), but more strongly compressed so that the rocks are tightly folded. If these imaginary rocks were subsequently planed off to a nearly, but not quite level surface, and the ground subsided, then the sea could invade. As the water rose, it would find its way first along the low-lying valleys, eroded along weaker rocks, while the harder rocks, forming ridges, would remain high and dry.

This is more or less how the Belcher Islands were formed, but with one interesting variation: the process described may be actually in reverse here. When the great continental ice-sheets lay on northern Canada, the weight of ice was sufficient to push the land below sea-level, perhaps by as much as 1000 m (3280 ft) around Hudson's Bay. Now that the ice has gone, the land is recovering slowly and heaving itself up, so that areas below sea-level a few thousand years ago are now emerging. Thus, the Belcher Islands resemble submarines surfacing infinitely slowly. Although no data are available for the islands themselves, it has been found that at Pelly Bay, just north of Hudson's Bay, the rate of uplift immediately after the end of the Ice Age was more than 12 cm (4.8 in) a year, although over the last few thousand years it has slowed to something of the order of 1 cm (0.4 in) per year. This rate will continue into the future for some time, so in a few centuries the Belcher Islands will have transformed themselves into a new, slightly bigger, equally enigmatic abstract painting.

Index

Image Dates

1. 27 June 1981;   2. 25 March 1975;   3. 20 January 1976;   4. 19 June 1975;   6. June 1983;   7. November 1981;   8. November 1981;   9. December 1983;   10. 2 September 1972;   11. August 1973;   12. 31 May 1977;   13. December 1983;   14. April 1983;   15. November 1981;   16. 23 December 1972;   17. July 1982;   18. December 1983;   19. 25 December 1972;   21. June 1983;   22. 10 September 1977;   23. December 1983;   24. December 1983;   28. December 1983;   29. 21 August 1975;   30. 30 April 1976;   31. 5 July 1975;   32. November 1982;   33. December 1983;   34. 28 January 1981;   35. 26 July 1973;   36. June 1983;   37. 4 October 1975;   38. 11 July 1975;   40. December 1983;   41. November 1981;   42. September 1983;   43. November 1981;   44. 23 November 1975;   46. November 1981;   47. 31 July 1973;   48. December 1983;   49. June 1983;   50. 22 June 1975;   51. September 1983;   52. 5 March 1976;   53. 5 September 1972;   54. December 1983;   55. 16 July 1973;   56. 10 August 1974;   57. 17 October 1968;   58. July 1969;   59. December 1983;   60. August 1973;   61. June 1983;   62. April 1981;   63. 7 October 1972;   64. November 1981;   65. April 1983;   66. June 1983;   67. November 1982;   68. November 1981;   69. June 1983;   70. December 1983;   71. June 1983;   72. August 1983;   73. December 1983;   74. December 1983;   75. June 1983;   76. April 1981;   77. November 1981;   78. December 1983.

The *Thematic Mapper* images (5, 25, 26, 27, 39, 45) are all more recent than 5 August, 1982, when the Thematic Mapper was first used.